James T. Turner, PhD
Michael G. Gelles, PsyD

Threat Assessment
A Risk Management Approach

Pre-publication
REVIEWS,
COMMENTARIES,
EVALUATIONS . . .

"**D**rs. Turner and Gelles have combined their research and experience to produce a book that sets a new standard for threat assessment professionals. *Threat Assessment* transcends risk management and should be read by human resource professionals, mental health practitioners, and the criminal justice community. Both authors were fortunate enough to work closely with Dr. Chris Hatcher and have done a magnificent job in representing Dr. Hatcher's assessment principles."

John A. Anderson
Chief of Police (Retired),
University of California;
Law Enforcement Training Coordinator,
San Diego Regional Training Center

"**D**rs. Gelles and Turner are both nationally known leaders in the field of threat assessment. Based on their extensive experience and comprehensive knowledge, they have now written an authoritative guide to the effective management of this growing and serious problem. It is a timely discussion that will inform professionals for years to come. If anyone wants to be conversant with these and related issues, they must be familiar with this book."

Melvin A. Gravitz, PhD, ABPP
Clinical Professor of Psychiatry
and Behavioral Sciences,
George Washington University,
Washington, DC

More pre-publication
REVIEWS, COMMENTARIES, EVALUATIONS . . .

"Human resources professionals and in-house counsel should utter a collective sigh of relief. Drs. James Turner and Michael Gelles have parlayed their vast years of experience with, and extensive study of, workplace violence into a step-by-step guide aptly named *Threat Assessment*.

For those on the front line in the workplace who know that the statistical odds are high that they will be forced to deal with a threat or incident of violence, this unique book methodically lays out the issues, the analyses, and the possible solutions. The case studies, based on thousands of actual cases on which Drs. Turner and Gelles consulted, are instructive and invaluable. No longer does a human resources professional have to freeze, not knowing whether she or he is facing a grave threat or simply a bluffing bully. Like conducting regular disaster or fire drills, reading this book provides the preparation one needs to calmly but rapidly access the threat, implement the crisis management plan, and, hopefully, avoid catastrophe."

Teri M. Solomon, Esq.
Senior Shareholder,
Littler Mendelson, P.C.

"This book is a must-have on the shelf of every leader not only for the techniques but as a guide to knowing whether you are getting good advice from your human resources and legal staff. These difficult situations, which all leaders will face in these times, can tarnish your leadership objectives unless handled with finesse and knowledge. Turner and Gelles have provided a structure and scripts that can be used to deal with a variety of crisis situations and to guide the organization's response in these challenging times. *Threat Assessment* is the resource to manage another type of chaos that can overwhelm business. These tools keep your people focused on your business and service objectives while contributing to the safety of all."

E. C. Murphy, PhD
Fellow, Murphy Leadership Institute;
Author, *Leading on the Edge of Chaos* and *Leadership IQ*

The Haworth Press®
New York • London • Oxford

Threat Assessment
A Risk Management Approach

THE HAWORTH PRESS
Risk Management
Joseph McCann
Senior Editor

Threats in Schools: A Practical Guide for Managing Violence
by Joseph T. McCann

Threat Assessment: A Risk Management Approach by James T.
Turner and Michael G. Gelles

Other titles of related interest:

*Kids Who Commit Adult Crimes: Serious Criminality by Juvenile
Offenders* by R. Barri Flowers

Violence As Seen Through a Prism of Color edited by Letha A.
(Lee) See

The Shaken Baby Syndrome: A Multidisciplinary Approach
edited by Stephen Lazoritz and Vincent J. Palusci

*Identifying Child Molesters: Preventing Child Sexual Abuse
by Recognizing the Patterns of the Offenders* by Carla van Dam

*From Hate Crimes to Human Rights: A Tribute to Matthew
Shepard* edited by Mary E. Swigonski, Robin S. Mama, and
Kelly Ward

*Political Violence and the Palestinian Family: Implications
for Mental Health and Well-Being* by Vivian Khamis

Program Evaluation and Family Violence Research edited by Sally
K. Ward and David Finkelhor

Risky Business: Managing Employee Violence in the Workplace
by Lynne Falkin McClure

Threat Assessment
A Risk Management Approach

James T. Turner, PhD
Michael G. Gelles, PsyD

The Haworth Press®
New York • London • Oxford

.

The Haworth Press, Inc., 10 Alice Street, Binghamton, NY 13904-1580.

PUBLISHER'S NOTE
Identities and circumstances of individuals discussed in this book have been changed to protect confidentiality.

Cover design by Lora Wiggins.

Library of Congress Cataloging-in-Publication Data

Turner, James T.
    Threat assessment : a risk management approach / James T. Turner, Michael G. Gelles.
        p. cm.
    Includes bibliographical references and index.
    ISBN 0-7890-1627-3 (hard : alk. paper)—ISBN 0-7890-1628-1 (soft : alk. paper)
    1. Risk management. 2. Threats—Prevention. 3. Violence in the workplace—Prevention. 4. Corporations—Security measures. 5. Behavioral assessment. I. Gelles, Michael G. H. II. Title.

HD61 .T875 2003
658.4'7—dc21
                                                                    2002027645

J. Chris Hatcher, PhD, died in 1999. He was the founder of National Assessment Services, Inc., a threat management group. Chris was our friend, our mentor, and is our co-author in spirit for this book. We worked together in the area of threat assessment in organizations since the late 1980s. Our thinking is permeated by our lengthy discussions together, the shared casework, and the training development. We have found it impossible to sort through and clearly define what is his and what is just the two of us. That is the magic of greater partnerships. The concepts and ideas are ours. Chris's insight and clarity of vision was to make threat assessment the best it could be. At times that could make him a challenge to work with. In the end, however, now that the last chapter of his life has finished and the book closed, we can say that he gave much more than he took. Chris, this book is our way of saying thank you for sharing with us.

ABOUT THE AUTHORS

James T. Turner, PhD, is President of International Assessment Services, Inc., a threat management company. He is also Senior Consultant and Principal for the Murphy Leadership Institute. Dr. Turner is involved in leadership development strategies and assessment approaches for senior leadership groups and boards. He has presented at a wide range of companies, performed hundreds of threat assessments, and trained threat management teams in a variety of organizational settings. He has appeared on "Nightline," "Marketwatch," and "This Morning," and has been quoted in the *Wall Street Journal, Fortune, Glamour, Health Care Risk Management,* the *VHA Alliance,* and *The Los Angeles Times.* Dr. Turner is a member of the Society of Human Resource Management and the Association of Threat Assessment Professionals. In 2001, he spoke on leadership issues at the National Association of Hispanic MBAs for the third year in a row and was also awarded the Hatcher Memorial Award for Visionary Presentations by the Society for Police and Criminal Psychology.

Michael G. Gelles, PsyD, is Senior Risk Assessment Professional with International Assessment Services, Inc., and Chief Psychologist for the Naval Criminal Investigative Service (NCIS). He assists NCIS and a multitude of other federal, state, and local law enforcement agencies with investigations and conducts psychological assessments of criminals and victims, as well as psychological autopsies. Dr. Gelles has conducted hundreds of risk assessments in the area of workplace violence, stalking, and threat behavior. He is an active member in a number of professional organizations, including the American Psychological Association, the International Association of Chiefs of Police, the Society for Police and Criminal Psychology, and the Association of Threat Assessment Professionals. He is a frequent lecturer, an editorial member of several journals, and the author of numerous published papers. Dr. Gelles holds appointments in psychiatry at the Uniformed Services University of the Health Sciences and at the Washington School of Psychiatry.

CONTENTS

Foreword

For decades, the nation has been experiencing the impact of workplace violence but has not been able to separate it from the violence that is a part of our society. Procedures for the prevention of workplace violence were rarely more than the security measures undertaken to protect against robbery and other forms of commercial crime. Many behavioral scientists theorized that containment of workplace violence required the same elusive solutions that would apply to the broad presence of violence in society. Fortunately, in the 1990s this changed with the pioneering efforts of a small group of professionals who studied how violence and the threat of violence occur in the workplace. Practical approaches and programs were created and introduced with substantial and measurable success. Although these efforts did not claim to transform violent individuals into pacifists, they did provide a special understanding of how the risk of violence can be assessed and reduced in the workplace. This process is demystified through the words and experiences of three veterans of the war on workplace violence. James Turner, Michael Gelles, and their legendary mentor, the late Chris Hatcher, are a new breed of psychologist who, along with a small number of other professionals, have brought structure and a beginning understanding to workplace risk assessment. It may be many decades before the same results can be reached across society, but in the workplace significant progress is possible and increasingly expected by the courts, administrative agencies, and most of all, the employees.

THE SIZE AND SCOPE OF THE WORKPLACE VIOLENCE CHALLENGE

The 1990s started with very little quantification of the workplace violence challenge. As attention focused on the workplace, largely because of dramatic incidents of violence, government agencies began to seriously collect information about homicide and other acts of violent crime in the workplace. According to the National Crime Victimization Survey (NCVS), 1.7 million violent crimes occurred in the workplace between 1993 and 1999. These crimes ranged from homicide to simple assault. The NCVS reported that 18 percent of all violent crimes could be classified as workplace violence (Duhart, 2001).

These statistics, while impressive, do not provide a comprehensive picture of the real problem of workplace violence. The National Institute for Occupational Safety and Health defines workplace violence as any physical assault, threatening behavior, or verbal abuse occurring in the work setting. This dramatically expands the scope of workplace violence. One study, a survey titled "Fear and Violence in the Workplace" (1993), identified threatened violence as impacting as many as one out of four employees. Understanding that threatened violence is a major part of workplace violence is essential to addressing this national challenge. Many times, the threat of violence is itself so damaging that it can exceed the harm caused by physical injuries. The psychological torment of a threat can last longer than the healing time required for many physical assaults. During the 1990s this understanding greatly increased, resulting in many states and the federal government adopting statutes defining certain threats as stand-alone criminal conduct. With the advent of the Internet age, the definition of prohibited threats often includes "cyberthreats."

THE RELATIONSHIP BETWEEN THREATS AND PHYSICAL VIOLENCE

The classification of threatening behavior and verbal abuse does not exist in isolation. In a majority of physical assaults that occur in the workplace, the perpetrator is someone known in the workplace. This could be a co-worker, a former employee, a customer, a vendor, a relative, or an individual with a romantic interest in someone within the workplace. During the 1980s, and more systematically in the 1990s, case histories show a clear pattern of verbal abuse, threatening behavior, and then escalating physical violence. This pattern has been consistently repeated. In the hundreds of cases this author has handled, when physical violence is caused by individuals known in the workplace, 99 percent of the time threatening behavior and verbal abuse preceded physical violence. At the same time, vast numbers of threats occurred without leading to physical violence. Accordingly, those responsible for workplace safety have faced the difficult challenge of attempting to determine the level of risk associated with threatening behavior and verbal abuse, knowing that in the majority of these instances, physical violence does not occur.

THE ESSENTIAL ESTABLISHMENT OF A MULTIDISCIPLINARY TEAM

This challenge is the focal point of this treatise. Turner and Gelles skillfully use and explain the tools of risk management pioneered by Chris Hatcher and refined through their own experience. At the center of this analysis is the recognition that the management of threats in the workplace must be undertaken by a multidisciplinary team including professionals from the fields of human resources, security, management, legal, and mental health, who are specially trained in threat assessment.

As with many challenges in society and the workplace, solutions are nearly impossible if the problem is seen from only one dimension. Workplace violence is the direct responsibility of security personnel. It is also an immediate concern for the manager of the threatened victim or alleged perpetrator. Prevention of violent behavior and uncovering the potential for violence are an important concern of senior management and an assigned job duty of human resource professionals. When the threatening party is a current or former employee, responses must be structured consistent with workplace policies and law. This involves input from the corporate legal department or an outside attorney schooled in the law of workplace violence. In the past ten years, an entire legal subspecialty has emerged dealing with workplace violence. Legal standards have started to develop for workplace safety, the duty to inform, responsibilities for protecting the rights of accused workers, negligent hiring, negligent supervision, privacy, protection of the disabled, and many other related topics. All of these team members must then integrate their knowledge and planning with the guidance of a psychologist or psychiatrist specially trained and experienced in dealing with such behavior. Successful management of the risk of workplace violence is dependent on the establishment of a team structure.

THE CHALLENGE OF PREDICTING VIOLENT BEHAVIOR

Once the organizational structure has been developed, then a critical portion of the management process is the risk analysis. The very concept of a risk analysis involving violent behavior engenders resistance. This comes from deeply held beliefs and legal principles that tell us that in a democracy people should not be singled out based on what "might" happen, but rather that adverse action requires actual provable misconduct. Profiling people for potential violence quickly invokes concerns of discrimination based on factors outside the control of the individuals such as race, national origin, sex, or even mental health. In a world where our genetic makeup can potentially predict future illnesses, great abuse is possible. Clearly, any form of discrimination based on these legally protected categories is not only wrong as a matter of law, but also wrong as a matter of prejudice. Our society has long rejected harming the livelihoods and careers of many to potentially reach the few who might actually engage in violence.

This overreaction is not necessary to accomplish the objective of managing risk. At first, when the previous concerns were combined with early studies that showed that even past violent criminal behavior was not necessarily a predictor of future violence, a useful risk analysis seemed unattainable. This environment confronted Chris Hatcher when he sought to find ways to prevent, or at least reduce, the likelihood of workplace violence. Hatcher's solutions came from his experience coupled with a genius for inserting a risk analysis into the context of a lawful organizational response to a known danger. These solutions are brilliantly reflected in the work and writing of his colleagues who have implemented his vision, expanded upon it, and seen it become the most widely accepted approach to combating the national problem of workplace violence.

Chris Hatcher was a superior public speaker and educator on the topic of workplace violence. One of the stories he often used to lighten the mood of the audience when introducing this serious subject was to refer to his grandfather or the grandfather of the person introducing him, whichever seemed least offensive. He then explained that this grandfather was well known in the Wild West. On one occasion, a group of soldiers encountered this famous frontiersman as they were looking for a wagon train they had been dispatched to protect. The grandfather was holding his head to the ground and informed the soldiers that the wagons had been through the area three hours earlier and had been attacked by six Native Americans. The driver of the wagon was wounded and the woman next to him was carrying a baby. Stunned by the detailed description, the captain leading the column asked how it was possible to know so many details merely by listening to the ground. The grandfather replied, "It's easy." Three hours ago, the wagon train rolled past him, and he was struck by the wagon driven by the wounded man. It then rolled over his head and left him in his current position.

When the audience predictably laughed, Hatcher explained that the study of workplace violence was very much like his lighthearted tale. Rather than making predictions based on some wizardry that would do the impossible, the process was very simple. For decades, incident after incident of violence in the workplace reoccurred with certain common characteristics. In many ways, he explained this was like getting hit in the head with the same information over and over. The task was to look at what was occurring, document it, and put it in language that could be understood. He would then give examples of cases with tragic outcomes that were preceded by threatening behavior and usually very clear verbal threats. To make the point, he would often cite a case that had occurred that day or within the last week. Frequently, a pattern of threats would be disclosed in the media coverage of the event. He then gave examples of incidents wherein the threats were not immediately reported, but after a period of time or in litigation, the pattern was again disclosed. Hatcher explained that often those who received the threats would feel guilty for not reporting them. In these cases, it might take some time, but eventually the disclosures were made.

These observations were not unique to Hatcher or the team of psychologists that he assembled into his national company, now called International Assessment Services, Inc. Almost every published paper or report on incidents of workplace violence involving known perpetrators shows the same information—a pattern of threatening behavior and verbal abuse prior to physical acts of violence. This is in stark contrast with people who rob or vandalize a workplace; they normally hide their identities. The more a perpetrator discloses his or her identity, the more likely a pattern of escalating activities can be shown.

CLASSIFICATION OF POTENTIAL PERPETRATORS BASED ON THEIR CONDUCT

Accepting that there is a repeated pattern of behavior associated with workplace violence, Hatcher was able to build a practical threat assessment process and risk management structure that is now being refined and implemented by his colleagues. After showing how information is carefully received and sought, the next stage in the risk assessment process is to use that information to place the alleged perpetrator into the Hatcher five-category system. These categories range from individuals falsely accused of threats or violence to those who are engaged in conduct that violates criminal statutes. In each instance, the classification is based on the conduct of the person being assessed, and then correlating that conduct with the reoccurring pattern described previously. The final placement requires the skill of a mental health professional, but at the same time all the members have a way of organizing and looking at the evidence. For example, a category three is a classification based on observed and reported conduct that shows actual threats and intimidation, but lacks evidence that the individual has or will engage in physical violence. Clearly, individuals could move from a category three to a category two if more is learned about their past or future conduct, such as actual violence against a spouse. Again, classification is based on conduct, not speculation or mere instinct.

Since the intake and interview process involves a report on actual conduct, it does not "profile" the subject. This is critical to minimizing the legal challenges to the actions taken by the organization, and keeps the focus on what occurred rather than what might occur. A person classified in category one would have engaged in conduct that violates criminal law and inevitably justifies a lawful separation from the work environment and the likely intervention of law enforcement. How this is accomplished would vary from case to case while the organization complied with its legally mandated duties.

In contrast to category one, categories two and three deal with conduct that likely does not violate criminal law but does violate the organization's policy on accepted conduct in the workplace. An isolated threat or one not correlated with other conduct associated with physically violent behavior is differentiated from threats made in context of other past and present conduct more associated with a pattern of behavior leading to violence. In each case, a termination or other disciplinary action in the workplace would be firmly grounded on the reported conduct, not the category in which the individual is placed or any prediction of future violence. This system recognizes that abusive behavior that threatens co-workers is itself misconduct that can justify discipline up to and including termination. The human resource professional and legal counsel are responsible to review the ability of the organization to act. Management then normally makes the decision regarding the appropriate action, which is then carried out with the assistance of security. If the conduct is sufficiently severe that it is associated with a greater likelihood of physical violence (a category two classification), then discipline (usually termination) will be carried out in a way that best protects the workplace and leads to a separation, while minimizing the likelihood that the organization's action will actually increase the potential for immediate violence.

This brief preview reveals the critical importance of the classification process in building a workplace prevention plan and strategy. Depending upon the classification, additional steps are taken to protect the workplace. However, in each situation, the actual conduct determines whether disciplinary action and/or involvement of law enforcement can be justified. In this manner profiling is avoided, even though it is possible to describe the likely characteristics of the group engaging in conduct that suggests an increased risk of violence. The great danger of a method that relies on profiling is that nonviolent individuals could be targeted based on a genetic condition that does not necessarily lead to violence. In addition, individuals might be over-looked because of a genetic characteristic (such as being female) when, in fact, the individuals' conduct suggests that they are engaging in an escalating set of behaviors that increases the risk of violence. Again, in each case under the methodology being presented in this treatise, risk management is based on provable conduct, not speculation or prediction.

THREATS ARE A FORM OF WORKPLACE VIOLENCE

Another obvious but critical element in the threat assessment process is the recognition that creating fear in co-workers is a form of violence. In managing risk, it is frequently sufficient to prove that a perpetrator engaged in intentional intimidation regardless of whether the threats are ever realized. This is prohibited conduct in the overwhelming majority of workplaces and alone justifies employer action. How that action is carried out and its magnitude might be impacted by the risk assessment process but, legally, discipline and limits are justified by the provable conduct of making the threat. Turner, Gelles, and Hatcher were among the first to draw attention to this type of activity as a workplace harm that justified a strong employer response. Indeed, as this has become more commonly understood across the nation, zero-tolerance policies have been widely implemented. Today, receiving a threat at work still occurs, as does the telling of a sexually offensive joke, but such events are far less common than just a decade ago. Clearly, the culture of acceptable conduct is changing, and the current authors can take pride in having been a part of the movement that has been the catalyst for this change.

DEMYSTIFYING THE THREAT ASSESSMENT PROCESS

One of the greatest contributions of this book is to demystify the threat assessment process. Too often, teams assigned to evaluate a workplace situation turn to the medical professional for a conclusion on the risk of violence. When this is done as though coming from a magical box, the team is deprived of vital information to support its action. Under such circumstances, current conduct can take a secondary position to the "prediction." This invites legal problems and more important, sets in motion responses to the individual that may not be in the best interest of protecting the workplace. Boldly, Turner and Gelles explain, in terms that all of us can understand, how this assessment process operates. This does not mean that the reader will become a qualified professional that can make such assessments, but because the conduct and process is understandable, the team can make better decisions on the appropriate preventive actions to be taken. If an employee makes a threat as part of a Halloween prank, the reported statement and the surrounding conduct of the individual may clearly indicate that neither violence nor intimidation was intended. This type of threat may violate an organization's zero-tolerance policy, but has far different implications than a similar threat made repeatedly and in the context of having committed past violence. Both individuals may be disciplined or warned for their misconduct, but the security precautions and monitoring for follow-up behavior would likely be very different—and should be dramatically different. Again, the category system signals this difference and helps the team decide upon the proper response.

This book puts in understandable words a process that has been tested hundreds of times in workplaces throughout the nation and beyond. It is mandatory reading for anyone responsibile for being part of a workplace violence assessment or response team. It has been said that the work of a true master is best seen in the paintings of his or her students. Clearly, the writings of Turner and Gelles are a masterpiece, made possible through the insight, courage, and teachings of Chris Hatcher.

THE LEGACY OF J. CHRISTOPHER HATCHER

The promise of this book and its teachings is demonstrated by turning the calendar back to the mid-1990s and witnessing the following situation that is based on real events. A well-known physician returned to his home after a full day in surgery. As he entered his kitchen a phone began to ring. He hesitated and then quickly answered, "Hello." Silence greeted him. "Hello, is anyone there?" A deep voice well known to the doctor finally responded, "Your time is coming!" The line went dead. This was the second unlisted number the doctor had switched to in the last six months. Nothing seemed to stop the former employee's calls, even though the physician had not seen him for nearly ten years. The doctor's face blushed, and his anger was only partially controlled as he called for his bodyguard. "Did you record the number he called from?"

No one was shot or physically injured, but for more than ten years, the prominent physician lived in fear of a former hospital employee. The hospital had provided the doctor with personal protection at his home as well as on the job. They knew the former employee lived out of state with his mother. They viewed him as a real threat, but dozens of times the hospital administrator had contacted law enforcement, private security professionals, and even attorneys all without results. It seemed that no one had an answer to a situation that did not show up on crime surveys or apparently fall under any available statute. Nonetheless, the harm was overwhelming, not just in security costs but also in imprisoning the doctor and his family in a world of fear.

This situation greeted Dr. Chris Hatcher and me a week after we appeared on an NBC news broadcast on workplace violence. Dr. Hatcher taught at the University of California, San Francisco (UCSF) Medical School and had dedicated his life to the study of violence and its prevention. He chronicled scores of cases and was frustrated by the lack of any organized processes for identifying and assessing workplace threats. Often, the assessment occurred only after the victim was physically injured or worse. Moreover, Hatcher recognized that a significant part of this national workplace violence problem was the fear and mental torture that came from threats. His teachings were combined with my knowledge and concern for the legal requirements of the workplace.

Within three weeks of being consulted, significant change was taking place in the physician's life. Under Hatcher's direction, the response to the threatening employee shifted 180 degrees. Instead of imprisoning the physician in a fortress and waiting for the next contact, the risk assessment process was directed outward toward the perpetrator. Two psychologists serving as consultants to the hospital interviewed him. He confirmed that he was responsible for the threatening behavior directed toward the doctor. Three hours were spent detailing his perceived grievances and, in many ways, invited the hospital to take action. He accepted that his statements could be used against him, and seemed to welcome a forum to reissue his threats. Contrary to the commonsense assumption that such a perpetrator would refuse to talk, he volunteered his personal history and details of his past employment. Accordingly, it was possible to cite specific conduct engaged in by the former employee and to make a threat risk assessment that suggested several avenues for containment. For the first time, it was possible to successfully involve federal law enforcement, seek the intervention of the courts, and redirect the focus of the perpetrator from the physician to the "governmental system" that enraged him. This in turn led to court-ordered treatment, the involvement of his mother, and a careful monitoring program.

For the first time in years, the physician and his family could again start enjoying their lives. A formal threat assessment had occurred, a detailed and individualized action plan was developed, and the focus of the former employee shifted away from the doctor. Tools and options were fully employed to give the doctor a sense of security that was impossible even with twenty-four-hour-a-day protection. The cost to the hospital for this service was a fraction of the monthly security costs. Although no one can provide guarantees, the threat assessment and the organizational response it supported gave the physician a new life. Not unlike a newly discovered antibiotic, the applied knowledge of Chris Hatcher and the employer's multidisciplinary response team had accomplished more in weeks than had previously occurred in years.

This is merely one of many cases that greatly benefited from Dr. Hatcher's work. So intense was the demand for his services that he formed what was originally known as National Assessment Services, Inc., and now has become International Assessment Services, Inc., under the leadership of James Turner and Michael Gelles. These professionals, in combination with a handpicked group of others from throughout the nation, became colleagues of Chris Hatcher, helping to develop and refine a process of threat assessment and risk management that has proved successful time after time.

By 1999 when Chris Hatcher was unexpectedly taken from us, corporate America combined with government was awakening to and embracing his message. Hatcher was the genius that had reached into the darkness and started to make sense out of the chaos. He advocated a multidisciplinary response to workplace violence and established a five-category approach for classifying threatening behavior. State legislatures, regulatory bodies responsible for workplace safety, and even the federal government modeled their workplace violence prevention programs based on Dr. Hatcher's teachings. The most used corporate educational video in the 1990s was "Workplace Violence: First Line of Defense." This teaching tool has been seen by millions and is still in wide use today. It features Chris Hatcher, his methodologies, and his practical approach to assessment and risk containment.

Throughout history many are remembered for their acts of bravery on the battlefield confronting violence with violence. However, in this post-9/11 world, another type of hero has gained our attention and deep appreciation. Firefighters, police, and other uniformed professionals who responded to the call for help make this a far better world, and have become role models for a society that someday may contain if not control the spread of violence. Chris Hatcher is the quintessential example of this new type of "hero." He dedicated his life to seeking answers for a problem that has been with us for centuries. He fashioned solutions at a time when none existed. He acted before all the verifications and separate studies would validate his methodology. He clearly saved hundreds from the pain of threatened violence and scores from the impact of physical violence. In so doing, he has given us a template that now defines mainstream workplace violence prevention policy. Tens of thousands now benefit from a workplace where the threat of violence is lessened, and the tools to respond to threats are now being implemented.

Chris Hatcher left behind a loved and cherished fifteen-year-old daughter who now can look forward to a society that offers a far better workplace environment due to the efforts of her father and those who now passionately carry and expand upon his mission.

Garry G. Mathiason
Senior Shareholder
Littler Mendelson, P.C.

Preface

In writing this book we determined its purpose to be one of support for the groups of individuals who struggle to create and maintain a safe organizational setting. This book is intended to assist these teams of human resource, legal, and management individuals whether they are called threat management, incident management, or crisis teams. We have learned that the assessment and management of threat behavior in the organization environment is truly a multidisciplinary industry. These professionals with proper training and support from consultants can responsibly and wisely manage a wide range of behavior. This represents a significant change from the mid-1990s when external groups directed and managed these problems.

Given the changes in recent legislation over the past several years in the area of harassment, communicated threats, telephonic threats, electronic harassment, and stalking, law enforcement can play a very different role in managing an employee or anyone else who communicates a threat or engages in behavior that poses a threat. Corporate America has increased its sensitivity to workplace violence and threat behaviors, actions, and statements. The number of policies in organizations related to situations that create domestic danger, fear, and emotional distress has increased. As a result, the workforce is more sensitive and reports incidents more freely than in the past. For the human resource professional, an increase in the number of cases reported places a greater demand on already taxed resources to manage threats. This requires an organizational system and methodology that ensures a reliable and consistent review in every case of a communicated threat.

The human resource professionals are on the front line in organizations when people problems arise. They are the first to receive the threat and, in essence, are the first to take some form of action to include reporting the incident up the chain of management. The human resource professional is the triage person who inevitably holds much of the early risk management responsibility and, in the long run, the liability for the company as it relates to the health and safety of its employees. Therefore, this book, in essence, is a guide for these frontline professionals. It outlines steps and strategies in which a communicated threat can be assessed and managed in a universally methodical fashion that attempts to organize an approach that reduces liability and enhances risk control.

We have found a great degree of secrecy among our associates in the threat assessment arena. The Association of Threat Assessment Professionals has done its part in trying to break down these walls of secrecy. In this book we share many of our core approaches to threat assessment in the hope that our action will encourage others to show detail of how they do what they do. Only by comparing and contrasting approaches and encouraging those with research roles to sort through the best efforts of those in the field can we continue to improve. Expanding the knowledge base through integrating research with the pragmatism and experience of those facing potential life-threatening problems can lead us to find more ways to protect the people who we try to assist while maintaining high quality organizational environments.

Life is too hard without any employee, customer, client, or manager having to wonder whether individuals meant what they said and whether he or she should go to work tomorrow. No individual should have to work in an environment where someone else's actions, statements, or behavior create fear. The organization clearly has a vested interest in building a safe, positive work environment while resolving conflicts and differences in perception.

Updates to techniques and tools shared in this book can be found at the Web site <www.threatassessors.com>.

Acknowledgments

This book is the result of years of casework and model building as the world has changed. We would like to acknowledge the contributions of our friends and associates from within International Assessment Services: Loren Brooks, Rick Bradstreet, Charles Burchell, Angela Donahue, Jim Janik, Gary Kaufman, Mike McMains, Kris Mohandie, and Gerald Sweet.

Cathleen Civiello, Neil Hibler, Joe Krofcheck, Michael Reynolds, and Tom Richards contributed to our thinking on insider threats. Kim Sasaki-Swindle contributed to interviewing concepts. Julian Hubbard contributed his advice, counsel, and good judgment as we explored concepts with him. We want to express our thanks for the professional contributions to our thinking that Robert Fein, Bryan Vossekuil, and John Monahan have made.

Special thanks go to two individuals, Garry Mathiason and E. C. Murphy, who offered unwavering support especially during the difficult time after Chris Hatcher's death. This was over and above the mentoring and sharing of their knowledge with us that had always been the case.

Beverly Maes was instrumental in the actual production of the manuscript. Her patience and long hours were a critical part of completing this project. Russ Palarea provided research support. Our families are due thanks for tolerating us during this process.

Finally, we thank Chris Hatcher for being so much a part of the work that we always have his counsel with us.

Chapter 1

Threat Management: Organizational Challenge

Threat assessment in organizational settings represents a unique set of challenges that require the merging of elements from behavioral sciences, labor law, law enforcement, and human resources. The purpose of developing such a model in an organizational setting is to guide the actions of management in responding to these situations in the workplace.

Management's goal is to provide the safest possible workplace while continuing the ability of the organization to sustain its mission. The objective of a risk management approach is to blend these areas of expertise into a plan that contains two core elements: (1) ongoing threat assessment, and (2) threat control action, threat management, and risk control.

In the case of truly random events, these actions may be futile in stopping an untoward event from happening. However, these types of events are relatively rare. The events that impact the functioning of organizations are often the result of a process that unfolds across time in a variety of settings and situations. These events can be confronted and responded to in an organized fashion. In doing so, organizations can manage risk to themselves, to others who work in the organization, and to their various stakeholders.

Threat assessment in organizations is more than focusing on overt acts of violence. It focuses on a variety of behaviors, actions, and/or statements that create intimidation and emotional distress in others, whether intended or not. These incidents are of concern to the organization because of their impact on the basic ability of the organization to deliver its legitimate set of services or the production of goods. The organization, in developing a risk management approach to threats, needs to consider the necessity of responding in a manner that allows employees and organizations to continue with their lives and daily routines. The response, or lack thereof, concerns not only safety but also the economic and service well-being of the organization.

As we have seen with the terrorist attacks of September 11, 2001, the impact of terror was not only on the direct targets. The collateral damage was inflicted on persons physically distant from the incident who watched or heard about what was happening. A second set of collateral impacts occurred to the economy of the nation brought on by uncertainty and fear. A primary goal of the administration in the early hours and days following the incident was to allay fear and emotional distress by giving a structured response to the national threat as well as by focusing on restarting economic activity. Each time an organization is faced with a potential threat or action, the leaders need to address the same issues on a smaller scale. By having a process in which employees have confidence, an organization reduces the ripple effect of such incidents. Failure to approach such incidents in a focused and organized manner, as we have seen repeatedly throughout corporate America, can worsen the situation. This poor response increases the risk to individuals while it creates distracting and costly legal problems for the organization.

The threat assessment approach views threats and overt acts of violence as occurring within a process or framework of variables that need to be understood and, where possible, managed. These situations can be highly fluid, with old information that is newly revealed, and new information and ongoing action by employees and managers that changes the overall assessment. Situations involving threat behavior or communications are not just "dangerous" or "not dangerous" but create for many employees emotional distress that is unacceptable and intolerable. These situations may progress over time from being perceived as benign to a potential for violence and physical harm. The goal of risk management is to reduce and contain the likelihood of not only physical violence but also intimidation and the intentional infliction of emotional distress. The threat assessment process is key to monitoring change or movement in a situation and the likelihood of increased or decreased risk for violence.

This emphasis on the risk management of threat moves away from older ideas of the prediction of danger toward the identification and management of a variety of communications, behaviors, and actions, as a way to reduce, contain, and manage the impact of such actions in the workplace. The process of threat assessment needs to reduce and contain the risk of inflaming the situation. This requires that management consider a range of actions from a variety of disciplines—labor law, law enforcement, human resources, management, security, and health care.

Heilbrun (1997) delineates the differences between prediction-based approaches and management-oriented approaches in conducting a risk assessment. The risk factors in violence prediction are static, individual-based variables, chosen to maximize predictions of the probability of a specific type of action occurring. In the management approach the risk factors are dynamic, related to the risk, and are chosen to be ones that can be impacted by intervention.

The risk management of threat behaviors has two intertwined primary components that we shall examine more closely: (1) a threat assessment process that is ongoing, and (2) risk control actions that are intended to manage the threat and contain/reduce the level of threat. If the actions in threat control are effective, the assessment process should reflect this and the threat assessment process should reflect changes in communications and behavior originally identified as creating emotional distress. Thus, the threat is assessed while it is managed.

In organizational environments, the ability to maintain focus on a single issue over a prolonged time period of even a few days is difficult. It is easy for the threat control steps to get lost, be overlooked, or not completed. This inaction or confusion can actually increase the risk and decrease the confidence of individuals in the organization in the threat response process. One of the key roles of those in threat assessment is to maintain the continuity of the process and to keep bringing the organization back to completion of those actions determined as necessary to manage and control the threat situation.

The goal of threat assessment is to move away from the prediction of danger or the imminence of violence toward the identification and management of risk. The model of action moves from a single individual making judgments on some set of empirical measures to that of a range of actions reviewed by professionals from a spectrum of expertise—labor law, human resources, security, management, and threat specialist. The ability to quickly assemble a team, develop a response structure, consider a variety of rapidly evolving pieces of information, structure a flexible action plan, and adapt that plan when information changes is needed by a threat management team. The multidisciplinary team needs to learn how to carry out designated acts within a limited time frame, and then regroup to assess the impact of those actions. This ability to assimilate and act on information is required to properly use the threat assessment and to initiate and maintain a response plan.

Threat assessment in this model requires several components, including a simple, straightforward manner of talking about risk that directs the organization into a structured response. Through the 9/11 tragedy we have learned that in times of emotional distress, leaders must respond rapidly, effectively, and with sensitivity, not only for those directly involved but also for other members of the community. The organization's leaders need to communicate to its members a sense of a steady hand on the wheel, a sense of knowledge and organization, and a sense of being able to move the organization in a direction to address the challenge. This is accomplished through corporate leaders who provide clear and concise information that creates the most realistic sense of safety and security. Without such a perception fear, panic, rumor, and inappropriate responses can lead to an atmosphere and actions that increase risk, enhance the likelihood of a negative outcome, disrupt ongoing operations, and increase emotional distress.

The goal then is to develop an approach to threat assessment that focuses on information that is legally available in the organizational setting and that incorporates ongoing reports of information as situations change. Management's response should focus on a highly targeted action, directed at the situation that is generating concern or distress. The response should include an overarching concern that lives may be at stake.

This approach emphasizes that we are looking at situations rather than just at an individual. The context and matrix of variables within which behavior, action, and communication occurs is the framework for risk management. In some cases, the alleged source of the problematic behavior may be unknown or the initial information unclear. Initial reports often provide a conflicting picture, with multiple players providing information that may have been known in the organization but never revealed. The catalyst is this new level of concern that causes management and co-workers to rethink their appraisal of past behavior and comments sometimes several years old and spanning multiple interactions.

Although this risk management approach shares many characteristics with the targeted violence approach (Fein, Vossekuil, and Holden, 1995; Fein and Vossekuil, 1998; Borum et al., 1999), some differences exist because of the nature of threat assessment in organizations. In organizations, the individual may be a part of the system, which creates a different framework of legal and human resources issues that influence the process of threat assessment. Also, the threats may constitute a wider range of concerns such as behavior that impairs the ability of co-workers to perform their legitimate job functions as a result of fear, intimidation, or loss of confidence in the ability of the organization to address issues related to their safety.

A number of principles as noted by Fein and Vossekuil (1998), anchor the development of a threat response model. Four of those are key to the risk management of threat in an organizational setting.

1. Threat behavior is the result of a process, not a random event. Different from targeted violence, there may be times when impulsive behavior(s) leads to the perception by others that a threat exists.
2. Threat behavior or the perception of threat occurs within a framework of behavior and circumstances that need to be reviewed and managed so as to reduce risk.
3. Review processes need to focus on making ongoing judgments as to the degree of risk that exists and what factors the organization needs to address in what order. This allows management to identify how these factors can be addressed with the resources that are available and what, if any, resources are needed from outside the organization.
4. Management plans need to be developed to direct the actions of individuals impacted by the situation, as well as communication strategies that conform to human resource concerns, organizational rules and policies, and state and federal statutes.

The skills necessary to manage these kinds of situations are often foreign to the organization. The need for rapid responses with limited information from multiple sources that may change over time often overwhelms the system's ability to cope and provokes a poorly considered, impulsive action by representatives of the organization that creates more risk and/or legal problems. Leaders need to put aside a "what we can't do" mentality and focus on the best solutions for managing risk, even when no perfect or optimal solutions exist. Systems are also taxed by the need to carry out the multiple response options in close sequence with a deadline looming.

It is 2:55 p.m. Within the last ten minutes one co-worker has come forward and said that last night John talked twice about being angry with his next two level managers and about going to buy an AK-47. John is due at work at 3:25 p.m. What should be done now?

A THREAT REVIEW TEAM

Because of the challenges outlined previously, organizations need a group of individuals who understand the process of threat assessment and who have experience in managing information, developing information sources, and using the expertise of threat assessment professions. A core team needs to consist of personnel from management, human resources, and security (internal/external), and people with expertise in labor law and threat assessment need to be available. The individuals chosen from each of these areas need to understand threat assessment and the interaction of this process with the different disciplines. Many general counsels provide an excellent sense of general law and a unique understanding of the organization dynamics; however, they may not have the specialized labor law expertise to provide advice on specific actions. Many mental health professionals have excellent knowledge of human behavior but little actual experience or knowledge of the specifics of threat situations and how these situations interface with other systems. Such can be said for each discipline, so it is important either to select individuals with specialized knowledge or to provide the opportunity for individuals within the organization to gain that knowledge. Such a team, in addition to having a good knowledge of the organization, may need knowledge specific to the industry (for example, nuclear extortion threats, celebrity stalking, and financial fraud). Identified experts who can provide input and updates quickly are also necessary.

The role of threat assessors requires a different set of skills from those required for other types of investigations and clinical/forensic settings (Fein, Vossekuil, and Holden, 1995). Actuarial methods that may be used with other populations (Monahan and Steadman, 1994) and to assess violence and mental disorders (Quinsey et al., 1998) are often not appropriate for populations in organizations.

The use of so-called profiles is extremely limited as they are often overly inclusive. A large number of individuals would fit the profile providing little useful differentiation. In our experience profiles lull people into ignoring or failing to respond to actions or comments by individuals who fall outside the profiles. Profiles disrupt and inhibit the assessment process. Threatening behavior or comments are the domain of no one type of individual. Again we emphasize that the context in which these actions or communications occur, rather than a set of characteristics specific to an individual within the context of the organization, is crucial to understanding and taking proper action. In threat assessment and risk control one seeks the commonalities in the process and the unique matrix of factors that are driving this particular problem.

For organizations striving to create a focused response to threat, a series of steps need to occur that allow them to develop an understanding of the context in which the communications or actions occur, to identify risk factors that may be influenced by some action, and to monitor the impact of ongoing intervention. To accomplish this an organization needs to develop:

- A team with specific skills
- A system for responding to the initial threat review process
- A shared model for conceptualizing these situations
- A set of actions that direct the management of risk factors

Chapter 2

Workplace Violence: An Overview

The phenomenon of violence has continued to receive increasing attention in our society. Such incidents have resulted in numerous legislative changes in the areas of harassment, terrorist threats, stalking, workplace violence, and domestic violence. Behavior that several years ago may have reflected poor judgment and aberrant conduct, today may be illegal and prosecutable. From the perspective of adjudication, behaviors that had been considered personal conflict may now be considered criminal.

Violence is best defined as a process. It is not an acute or episodic event, but rather an action resulting from a culmination of a series of failures, disappointments, and problems (Fein, Vossekuil, and Holden, 1995). The result can be an expressive, destructive, volatile behavioral outburst with the potential for grave consequences. Violent acts are the result of a process that can follow a patterned continuum. People who make verbal threats do not always pose actual physical threats (Fein and Vossekuil, 1998; Fein, Vossekuil, and Holden, 1995). The continuum starts with ideas and ends in action. Other individuals may move along the continuum from having such ideas to expressing them and then to engaging in behavior that rapidly becomes a problem. For those who tend to overtly express their anger and frustration, violence can be perceived as a solution to their problems.

Although there is no particular profile of these individuals, a sequence of identifiable behaviors illustrates this continuum (Fein and Vossekuil, 1998): After a conflict or problem the ideation of violence is perceived as a possible solution to the perceived stress or conflict. An individual may express this in thought only or may engage in some behavior that reflects potential future action. In many cases, individuals never go beyond the thought. At the other end of the continuum are individuals who communicate very little and engage in behavior that poses a serious threat to others. An individual may talk about assaulting someone and never make an approach, or an individual may appear angry and disgruntled and assault someone without communicating the intent to do so.

Violence is the result of a dynamic process consisting of an interaction of several factors (Fein, Vossekuil, and Holden, 1995). One set of factors is those characteristics inherent to the individual. This may include a history of impulsivity and a previous problem with violence in which the crucial factor is the outcome that resulted from the violent act. If an individual has used violence in the past, and has suffered few if any consequences, the actions are reinforced. The individual may perceive that violent acts are the preferred solution to problems. A sequence of behaviors, thoughts, and perceptions can serve as a trigger for the individual who perceives life's problems as essentially unsolvable based on experience. Although the problem under review may just be another in a series of problems, it becomes a trigger event. For the individual, this problem has triggered some movement from idea to action. Finally, individuals may be in settings that facilitate or do not inhibit or impede violent acts from occurring in the community, the workplace, or the home.

Other critical factors (Hatcher, personal communication, 1996) that contribute to such situations are:

- Individuals begin to blame or project responsibility for their problems onto the organization or an individual.
- Individuals begin to perceive the organization's actions as being directed against them.
- Individuals shift from a perspective of self-protection to self-preservation.

These factors may lead the individual to engage in both threatening and intimidating behavior to create distance from the perceived threat or to initiate a violent act to eliminate the perceived danger.

DEFINING WORKPLACE VIOLENCE

Workplace violence has become a familiar term in the United States over the past ten years. Although the annual statistic for workplace homicide is static (Duhart, 2001), the number of assaults and threats is exceedingly high. In 1998, the Office of Personnel Management published a manual titled *Workplace Violence for the Agency Planner.* This publication followed a significant change in corporate America, where many companies and corporations were developing specific policies regarding workplace safety. Most of these policies address, define, and frame acceptable behavior in the workplace. In many cases, violation of these policies results in progressive discipline, a referral for counseling to the company's employee assistance program, or termination of an employee.

For many years workplace violence was considered a boutique crime. Beginning in the 1980s workplace violence was primarily managed by corporate or private security. In many instances, problem employees were either transferred or ignored when their behavior in the workplace was perceived to be threatening. In the early 1990s many company managers began to view workplace violence as a significant issue that required greater attention and a more multidisciplinary approach. During this time, legislative changes were occurring in the United States pertaining to occupation health and safety, stalking, and targeted violence. Law enforcement began to take on a more central role with private security in the forefront of investigations of communicated threats and violence in the workplace.

Today, many changes to laws address harassment, communicated threats, stalking, and domestic violence. Much of the new legislation covers threatening communication or behavior in the workplace. Private and corporate securities have developed new roles as multidisciplinary partners with human resource professionals and management to conduct thorough, comprehensive investigations and assessments of possible violent behavior in the workplace. The increased attention to a safe workplace has resulted in greater sensitivity and a wider array of management responses to threatening communications and behavior. However, despite the past ten years of increased attention and vigilance in much of corporate America to workplace behavior, company managers still remain reticent in many cases to more aggressively confront inappropriate and emotionally distressing behavior on behalf of co-workers and other stakeholders.

In spite of a company's effort to develop, publish, and enforce a workplace violence policy, it continues to be easier to delay dealing with a problem employee's behavior or to transfer the employee. Although many companies have a workplace violence policy and espouse a zero-tolerance attitude regarding workplace violence, they often do not have an organized plan or method in which to address communicated threats and frightening behavior in the workplace.

Behavior that caused emotional distress for employees or appears to be unacceptable in the workplace requires a multidisciplinary approach. As stated previously, these situations require a review by personnel from management, human resources, corporate security, internal or external legal counsel, and a psychologist who has experience in risk assessment.

Corporate employee assistance programs have a role to support employees who demonstrate problem behavior in the workplace. The expectation for confidentiality by the employee who enters the employee assistance program process will directly limit the amount of information management may access in making an employment decision. Organizations need to protect the credibility of the employee assistance programs; such programs have worked hard to establish themselves with employee groups. Therefore, it is seldom advisable for employee assistance representatives to be core members of threat assessment teams. They can best serve as a resource without perceived conflicts of interest by being available to all employees, including the employees of concern.

Management requires information about an individual's behavior in the workplace when the employee has been reported as engaging in threatening communications or behavior. The work-related review focuses on an employee's behavior in the workplace as observed by co-workers and supervisors. A work-related behavioral review as part of the risk management provides management, human resources, security, and legal counsel valuable insight into the situational matrix of variables that have generated concern, including actions and/or words that have become a source of emotional distress for co-workers.

RECOGNIZING THE TYPES OF WORKPLACE VIOLENCE

There are several types of workplace violence. An individual perpetrates the first type of workplace violence with no legitimate business interest in the company (e.g., an individual who robs a taxi driver or liquor store). Such individuals are engaged in behavior to serve their own needs that may be separate from any previous relationship with the organization or in response to services or products. This occurs in approximately 60 percent of the documented cases. A customer or service recipient of the company perpetrates the second type of workplace violence (e.g., the individual is dissatisfied with a service or product). This occurs in about 30 percent of the cases. Finally, someone with an employment-related relationship perpetrates the third type of workplace violence (e.g., disgruntled employees who were either recently terminated or failed to receive something they felt entitled to). This occurs in about 10 percent of the cases (BLS, 1995; Castillo, 1994; Jenkins, 1994; NIOSH, 1992, 1993, 1995; Windau and Toscano, 1994). The following are case examples of the different types:

- *Type 1:* A retail shop owner receives a gunshot wound to the shoulder during a robbery by a person who was unknown to him and who had no current or previous relationship with the store.
- *Type 2:* A client of a financial brokerage house enters the office and attacks a financial planner with a club. The attacker claims that the brokerage house mismanaged his money and was responsible for his recent bankruptcy.
- *Type 3:* A recently terminated employee returns to his office and assaults his former supervisor. The employee claims that he was unfairly terminated and that the supervisor had it in for him for several months.

STATISTICS ON WORKPLACE VIOLENCE

Workplace violence is the leading cause of death for women in the workplace and second leading cause of death for men. Men are three times more at risk than women for workplace homicide. Workplace homicide remains a fairly stable statistic. There appears to be about 900 to 1,000 deaths nationwide. There are an estimated twenty workers murdered each week at work. Overall, these statistics are considered to be underreported by 25 to 50 percent (Jenkins, 1996; NIOSH, 1992).

In addition to the workplace homicide statistics that remain fairly constant, some of the other statistics concerning workplace violence are staggering. For example, 1 million Americans are attacked, 6 million are threatened, and 16 million are harassed on the job each year. Estimates indicate that 18,000 workers are assaulted each week. Workers in health care, community service, and retail are at increased risk of nonfatal assaults. Statistically, workplace assaults occur in about 61 percent of private employment, 30 percent among government employees, and 8 percent of the self-employed. Fifty-six percent of assaults are believed not to be reported. The retail and service industries account for more than half of the homicides, as well as 85 percent of nonfatal assaults. The cost of this phenomenon in dollars to industry is staggering—in the millions of dollars annually. This cost is the result of the loss of productivity, work disruptions, employee turnovers, litigation, and other incident-related costs (Duhart, 2001).

Workplace Guidelines

Workplace violence has spawned an increasing number of legal mandates. These began with the Occupational Safety and Health Administration's 1970 general duty clause—"duty to provide a safe work environment." In 1994/1995 workplace security-required programs were revised and included additional guidelines regarding hiring, training, supervision, and retention. These guidelines call for preventing harassment and discrimination. The guidelines also emphasize the duty to warn intended victims whenever a person is the target of a communicated threat. The guidelines protect the employee by outlining parameters regarding defamation, wrongful discharge, and invasion of privacy. All of the guidelines appear to be in accordance with the Americans with Disabilities Act at the time.

MYTHS REGARDING WORKPLACE VIOLENCE

A number of myths regarding workplace violence highlight some of the misconceptions about this phenomenon:

Myth: *There is a demographic profile of the potentially violent employee/individual.*

Reality: Perpetrators of workplace violence have many different demographic backgrounds.

Reality: Demographic profiles may cause us to ignore potential threats.

Reality: Behavioral clues/indicators are more important.

Myth: *The workplace violence problem is all about homicide.*

Reality: The workplace violence problem is less about homicide and more about assaults, intimidation, and fear in the workplace, which occur far more frequently.

Myth: *Workplace violence is most typically committed by a disgruntled employee.*

Reality: There are three types of workplace violence perpetrators—someone who has no relationship with the company, a customer or service recipient, and someone with an employment-related relationship with the company.

Reality: The disgruntled employee is the least frequent, but most publicized, source of incidents.

Myth: *Individuals are dangerous or not dangerous.*

Reality: Individuals may fall along a continuum of violence potential and risk for creating emotional distress in co-workers.

Myth: *Violent employees strike without warning or clues.*

Reality: Potentially dangerous individuals present multiple clues to multiple people prior to an incident.

Reality: Clues are not reported to appropriate individuals.

Reality: Clues are verbal (sometimes bizarre) statements, physical acts, and obsessions.

Myth: *There is no way to predict employee violence.*

Reality: Clues to employee violence are present but often are not relayed to managers and others who need to know.

Myth: *Violent employees have major job and personal losses.*

Reality: Violent employees have a degree of success.

Reality: Violent employees believe that they are morally entitled to something/someone that the organization and/or someone in the organization has taken away or prevented them from having.

Myth: *Even if you can identify violent employees, you cannot do anything about them.*

Reality: Intervention by boundary control and team management works when the threat assessment team:

- Achieves team agreement on management plans,
- Implements consistent communications to the threatening individuals, and
- Sets and maintains clear boundaries for the threatening individuals.

CONDITIONS THAT MAY LEAD TO WORKPLACE VIOLENCE

A number of conditions contribute to workplace violence. They include multiple economic factors, a rise in temporary and part-time workers, widespread downsizing past and present, and popular media factors including movies, games "to kill your boss," and comics/cartoons.

Certain conditions may lead employees to feel angry. Some employees feel angry when threatened with job loss. Others have a sense of entitlement to lifetime employment. They often believe that they are more capable than those being promoted or who survive layoffs. They expect that as long-term employees they have "prior rights" and certain status. Individuals may, because of economic conditions, have been through a series of layoffs with different organizations or within the same organization. These situations may lead individuals to engage in behaviors, actions, or communications that serve as warning signs.

WARNING SIGNS

Numerous lists of workplace violence warning signs have been published by a variety of sources. We have organized the more significant indicators into specific areas that assist in evaluating behavior and communications from a "whole person" model perspective.

Workplace Violence Indicators

Verbal Clues

- Direct and indirect threats
- Threatening/harassing phone calls
- Recurrent suicide threats or actions
- Hopelessness
- Boasts of violent behavior or fantasies
- Frequent profanity
- Belligerence
- Challenging or intimidating statements

Bizarre Thoughts

- Paranoid
- Persecutory delusions with self as victim
- Delusions in general
- Command hallucinations
- Significantly deteriorated thought process
- Obsessions
- Signs or history of substance use/abuse

Behavioral Clues

- Physical altercation/assault upon another person
- Inappropriate weapon possession or use
- Physical intimidation
- Following and surveillance of targeted individual
- Short-fused, loss of emotional control, impulsive
- Destruction of property
- Deteriorating physical appearance and self-care
- Inappropriate displays of emotion
- Isolated and withdrawn

Obsessions

- Self as victim of a particular individual
- Grudges and deep resentments
- Particular object of desire

- Perceived injustices, humiliations, disrespect
- Narrow focus—"sees no way out"—"no other options"
- Publicized acts of violence
- Weapons and destruction
- Fairness
- Grievances and lawsuits

Chapter 3

Organizational Response Options

As manager of operations you are sitting at your desk at 2:00 p.m worrying about the continued decline in orders. You are not sure how long production levels can continue as they are. Just then you hear a knock at your door. It's Tom, an average employee whom you have known for two years. Tom asks to speak with you. "I am not sure it's really anything but I thought about it overnight and talked with my wife. Now, I don't want to get anyone in trouble but I just thought you should know." "Know what?" you ask with a sinking feeling in the pit of your stomach. Just what you don't need. "Well," says Tom, "I worked the swing shift yesterday and, you know, everyone is talking about whether there is going to be a layoff. Well, Bob started talking to a couple of us and said if he didn't get a good package, he was gonna set things straight. Then he started talking about buying an AK. I don't think it means anything, but I thought you should know."

The situation now begins. In many cases, statements such as these will not be reported, especially where reporting policies are not clear and are not reviewed regularly with employees. Tom will try to make sense of the conversation, or he will dismiss it until some other action raises concern, or he will decide it will cause too much trouble or risk for himself to report it.

Monica, the human resources director of an East Coast building construction firm, receives a phone call. "Monica, this is Hosea, construction manager at the West Coast site. I'm not sure exactly what to do so I called you. Over the last week, we have found a series of messages in the outdoor toilets at the site. At first we just thought it was a joke and laughed it off. Today, however, the message really concerned me. It says that in three days, on the twenty-first, he is going to start shooting people working at the site."

Concerns come from far-flung parts of the organization only after people have been trying to deal with the situation, ignoring it, passing it off as a joke, or discounting it because, "It's just John." Managers are often asked to address situations across multiple time zones with minimal on-site presence.

At times such as these an organizational response process is important on a number of levels. It is important to safety, morale, potential legal exposure, and to minimizing the potential for disruption of the organization's work. The reactions of managers are critical in making sure that the content is forwarded to human resources or the designated department. For this to occur consistently, the middle level of the organization needs to feel confident that the threat response process will be neither an over- nor underreaction. Without confidence in the process, supervisors and managers will often attempt to resolve these situations on their own, and the problems will spin out of control. The culture of the organization needs to support managers who seek assistance with these issues and view such behavior as a characteristic of maturity, not inadequacy.

THREAT RESPONSE PROCESS

Each organization's leaders need to decide whether a situation triggers the threshold of the threat response process or whether the supervisor should be asked to informally talk with employees and try to gain additional information/resolution. If the situation triggers the organizational threshold for a threat response, a series of steps needs to occur.

Stage 1

1. Describe the threat/threatening behavior
2. Determine, with limited data, if immediate action (police call) versus prompt action (threat management team review) is necessary
3. Determine if a general threat or specific threat to person/property exists
4. Determine who heard/observed/received the perceived threat/threatening behavior
5. Determine if the team or manager has any knowledge of prior threat, threatening communication, angry outburst, or violence

6. Determine the level of distress by persons receiving, observing, or being targeted
7. Determine if the threat requires immediate security presence
8. Determine if the situation warrants the removal of any employees from the work environment
9. Review previous decisions as to whether the threat or action requires immediate police response
10. Access availability of external consultants

During this initial stage, basic fact finding needs to occur and some decisions need to be made to secure the safety of the environment and the individuals. It is important in steps 1, 3, and 4 to be as precise as possible.

Stage 2

Review incident and management decisions to date on a conference call with labor counsel and threat assessment consultant(s).

Although labor counsel and the threat assessment professional should be involved as quickly as possible, Stage 1 review by some members of the response team will many times be necessary without them due to the needed speed of response and the unavailability of the consultants.

One member should be designated to provide the situational review on the conference call. The information presented should be scripted and agreed to by the core team. In training threat management teams, we have seen numerous cases where, without scripting, the team or a member of the team gave faulty information to the consultants and then received good but wrong advice. This advice was based on incomplete or distorted information. At other times, teams try to present by sharing their impressions with little structure, and the labor or threat professional must slow the process to make sense of the confused presentation. This can lead to poor outcomes because of erroneous or absent information that would have caused the consultant to take a different direction. It also slows the response time and can increase consulting costs.

Stage 3

Identify information needed for further response:

1. Employees to be interviewed and by whom
2. Nonemployees to be interviewed and by whom
3. Background checks (consider with legal input)
4. Law enforcement liaison initiated as appropriate and by whom

Stage 4

Review site security management:

1. Additional information, if any, incorporated into threat assessment
2. Changes in level/type of security considered/contacts initiated to determine resource availability

Stage 5

Determine type of new case information, threats, or behavior that would trigger reevaluation of the level of security.

Stage 6

Determine threatened or at-risk individuals and/or sites:

1. Identify content of corporate communication to threatened or at-risk individuals/sites
2. Identify content of corporate communication to threatened or at-risk nonemployees (vendors, customers, etc.)
3. Identify who will conduct notification and coordinate with communication department if one exists
4. Identify counseling resources (employee assistance program) if needed by threatened or at-risk employees/sites

Stage 7

Determine need for interview of employee alleged to have engaged in threatening behavior:

1. Assess need to determine ongoing risk or overt behavior
2. Assess need to determine threat/creation of emotional distress

Stage 8

Determine who will conduct employee-in-question interview:

1. Internal resources based on experience, training, and extended availability
2. External risk assessment consultant

Stage 9

Determine risk level and response:

1. Internal model for categorizing risk
2. Consultant-provided model

Stage 10

Review of information available as a result of threat investigation:

1. Include labor law counsel, threat assessment team, threat assessment professional
2. Review data as well as any response to risk-control measures instituted
3. Threat assessment professional should not be present during deliberations that relate to employment status. Threat assessment is not part of employment status decision making. Threat assessment professional can be consulted after a decision has been made on employment status as to risk reduction strategies in response to the employment decision.

Stage 11

Determine civil/criminal law options.

Stage 12

On-site security responses should be coordinated as needed by security staff.

Stage 13

Off-site security responses should be coordinated by security staff as needed.

Stage 14

Track status of individuals impacted.

INCIDENT RESPONSE CHECKLIST

The following is a checklist of actions that threat assessment teams need to track as they proceed through a situation. This is a general guide, which must be adapted to individual situations in consultation with labor law counsel.*

1. Initial management team consultation (by management team, labor law counsel, risk assessment staff, security staff)

Team Members:

HR Staff Member _____

HR Staff Member Phone _____

Security Staff Member _____

Security Staff Member Phone _____

Legal Staff Member _____

Legal Staff Member Phone _____

Risk Assessment Team _____

Phones _____

Pagers _____

2. Immediate action as needed (by management team)
Remove employee from work site; conditions [administrative, medical, pay, no pay, restricted on-site access, restricted phone access, etc.]

Completed? Yes/No

---

*Copyright 1999, James T. Turner.

3. Contact with labor law counsel as needed (by threat team)

Completed? Yes/No

4. Contact with risk assessment professional (by threat team)

Completed? Yes/No

- Seek initial assessment of situation stability/areas overlooked/ areas needing additional inquiry
- Seek options for obtaining further information
- Determine conditions for further interviews to be conducted by management and/or security professional versus conditions for further interviews to be conducted by risk assessment specialist

  Option A: Risk assessment specialist assists in scripting interviews for company personnel

  Option B: Risk assessment specialist conduct interviews

  Option C: No further action at present; establish threshold for risk assessment specialist interview of co-workers/employee of concern

5. Background check as needed (by security staff)

Legal OK? Yes/No
Completed? Yes/No

Results _____

_____

6. On-site preinterview security enhancement as needed (by security staff)

Personnel Needed? Yes/No
Personnel Assigned? Yes/No

7. Off-site employee security response as needed (by security staff)

Need to restrict access to site? Yes/No
Need to restrict access to co-workers? Yes/No

8. Law enforcement agency liaison as indicated (by internal/external security staff)

Completed? Yes/No

Name of command officer at law enforcement agency_____

Is command officer to be on duty at time of interview?

Yes/No

Prepare briefing card with specifics of employee of concern?

Yes/No

9. On-site interviews of co-workers
   Name: _____
   Name: _____
   Name:_____

   • Are the above co-workers scheduled to work on the day of the interview?

   Yes/No

   • Are the above co-workers entitled to union representation?

   Yes/No

   • If necessary, how will the union be notified and by whom?

   _____

   • Will notification of the above co-workers about the interview be done by memo or verbally? _____

   _____

   • Is there agreement about the messages to be conveyed in the notification of co-workers?

   Yes/No

10. Management of target (at-risk) employees (by management and security staff)

   • Who is the designated management liaison to targeted (at-risk) employees? _____

   _____

- Will there be security considerations for such employees at the work site (e.g., parking place close to building, security escort to/from parking, cell phone loan, etc.)? _____

- Will there be security considerations for the targeted (at-risk) employees away from the work site (time off, visit friend/family out of area, police liaison)? _____

- What contact, if any, will be maintained with the spouse/relatives/partner of at-risk employee? _____

- If the employee is referred to employee assistance program (EAP), who will contact EAP manager to ensure personalized referral and briefing of designated provider? _____

- If the employee intends to utilize mental health provider from health plan, who will contact and provide briefing to the provider with the approval of the employee? _____

11. On-site interview of potentially violent employee (by risk assessment staff or internal resource with security backup)

    Name: _____

    Employee number, if applicable: _____

    - Is the above employee scheduled to work on the day of the interview?
      Yes/No

    - Is the above employee entitled to union representation?
      Yes/No

    - If necessary, how will the union be notified and by whom?___

    - Will notification of the above employee about the interview be done by memo or verbally? _____

- Is there agreement about the messages to be conveyed in the notification of the employee?_____

  _____

- Which management team member makes notation of employee agreement to come to interview?_____

  _____

- Which management team member will greet the employee at time of interview and introduce interviewers?_____

  _____

- Management team member needs to inform employee of the following:
  __This interview is an internal review of problem behavior between the employee and others
  __This interview is nonconfidential
  __All information gathered may be provided to management
  __The employee has access to the management team member during the interview if necessary at phone number____

  _____

  __The management team member will meet with the employee after the interview in person or by telephone, if appropriate
  __The employee can take breaks at any time
  __The employee can consult (if applicable) with a union representative at any time
  __Refreshments (sodas/coffee/snacks, etc.) are available in the room
- Interviewers make notations of employee willingness to proceed with the interview and obtain signature on interview sheet

12. Assessment of violence/threat categories 1 through 5 (by risk assessment staff)

    Category 1: Available data suggest high violence potential; behavior suggests immediate arrest, hospitalization, or ma-

jor organizational response. What will be the anticipated management response if the situation appears to be a category 1?

Category 2: Available data suggest high violence potential; behavior does not suggest immediate arrest, hospitalization, or major organizational response. What will be the anticipated management response if the situation appears to be a category 2?

Category 3: Available data suggest insufficient evidence for violence potential; sufficient data support perception of knowing infliction of distress on co-workers/others. What will be the anticipated management response if the situation appears to be a category 3?

Category 4: Available data suggest insufficient evidence for violence potential; sufficient data support perception of unplanned infliction of distress upon co-workers/others. What will be the anticipated management response if the situation appears to be a category 4?

Category 5: Available data suggest insufficient data for violence potential; insufficient evidence of behavior resulting in distress in co-workers/others. What will be the anticipated management response if the situation appears to be a category 5?

13. Post-risk assessment interview feedback session (by risk assessment professional(s) to management team, legal counsel, security)

   • What is the date, time, and location of the interview feedback session? _____

   • Who will be involved in the interview feedback session? ____

   _____

   • Who (threatened employees, witness employees, general employee population, public contact employees, etc.) beyond the management team needs to know what (notification for safety versus privacy issues) from the interview results? ____

   _____

14. Post-risk assessment interview on-site security enhancement as needed (by management team and security staff)

Personnel Needed? Yes/No

Personnel Assigned? Yes/No

- Is a restraining order warranted?

Yes/No

- What boundary-probing behaviors should be looked for? ___

  _____

- What boundary-probing behaviors warrant considering further contact (e.g., field interview) with the individual of concern? _____

  _____

- What boundary-probing behaviors warrant considering law enforcement action? _____

  _____

15. Post-risk assessment interview personnel action (by management team with legal counsel as required, risk assessment staff do not participate, should be absent)
Behaviors upon which personnel action is based? _____

  _____

Chapter 4

Intake Process

The intake process is the first step in any consultation. In many cases, it may be the first time that management has used an internal/external risk consultant to manage a potential threat incident. The threat assessment team needs to recognize that management, for all intents and purposes, is in crisis and requires an immediate intervention that gives some format or structure to an anxiety-producing and disruptive situation. Bringing an air of confidence and experience to the situation assists management in relaying crucial information regarding an individual and the perceived risk to employees or stakeholders that may exist in the workplace. Management is seeking immediate guidance and direction in how to handle the current employment problem. Managers, while not necessarily looking for an outside resource to make an internal management decision, are looking for potential solutions to a problem that they do not confront on a daily basis. They are looking for a consultant who is able to listen closely to the information available, provide some containment to a crisis situation, and offer direction as to how to develop strategies that will lead to a resolution.

INITIAL STAGE OF CONTACT

The initial communication with management can be the most critical of all communications in a case consultaion. The manner in which the information is conveyed to the consultant is crucial to the initial ideas and opinions formulated in an assessment. The assessor can only react to the information made available; therefore, capturing and organizing critical information in the early stages of an assessment can be the key to a quick and safe resolution to a workplace situation.

In many cases, the employee who engages in threatening communications or behavior has done so for a period of time. Although that time period may be unclear, engaging in threatening, intimidating communications or behavior is often the end point along a continuum from idea to action. Such behavior does not come to the attention of management until the individual has engaged in a series of often increasingly disturbing behavior either toward or around others in the workplace.

The consultant will receive a call from management that has a perceived crisis. Although the issue may now present as a crisis, it is likely that the problem has been ongoing over the course of an undefined period of time. It is important to assess early what exactly occurred that motivated the organization to contact the consultant at this particular time. The informant may find it difficult to attend to all of the necessary data that a consultant might find critical in making an assessment regarding the threat potential, so structure and clarity are important.

The intake procedure assists management in relating the necessary information regarding a problem to the assessor. The process can also provide a structure that management finds reassuring.

When initially discussing a case with management, the consultant should determine if the client has previously worked with the consultant on other employment issues, participated in training provided by the consultant, and/or received training or consultation from other professionals in the threat assessment field. The consultants need to briefly educate the client on how consultations are conducted and provide a short review of their philosophy regarding the assessment of threat behavior in the workplace. Since different professional disciplines have different approaches to threat assessment, it is important to explain the approach that will be taken in this case. For example, a psychologist may be behavior based while a security professional may use a safety-based perspective.

Initial Call: New Client

The following illustrates how IAS consultants handle new clients.

1. Introduction: "This is Dr. Robert Smith. I am returning the call you placed to International Assessment Services regarding a threat as-

sessment. By way of introduction, IAS is a company that provides assistance in difficult personnel situations. We conduct risk/threat assessments. We have multiple locations around the country from Los Angeles and San Francisco to New York and Atlanta. We are a twenty-four-hour-a-day/seven-day-a-week company and we can respond on site to diverse geographic locations. Our referrals come from employment law attorneys, human resource managers, and security directors. We do not work for individuals, only for organizations. By this time you are probably dealing with a situation that includes multiple individuals and specifics; therefore, let's not speak in general terms." Since initial calls are frequently unpredictable, the consultant may inform the caller how much time has been allocated for them at this time. If delays prevent starting the call, it may be necessary to reschedule or be referred to another consultant.

2. The consultant will inform the caller how the consulting group works. Initial consultations are typically per hour, rounded to the nearest hour, with a one-hour minimum.

3. Proceed with the call.
   a. Encourage the use of first names.
   b. Offer to fax or e-mail information about one another, including a services agreement, if needed.

4. The initial call should be closed on a positive note.
   a. The consultant may suggest that the threat management team have an internal huddle and then call back.
   b. The consultant will close the call with the following summary: "You have placed this call to seek the assistance of a specialized resource. This is the type of situation that involves a very small number of all employees. We've talked about a structure for understanding this situation and a strategy for proceeding with an assessment and developing risk control options. While you are in the process of deciding what to do, the employee in question may escalate his or her problematic behavior. If so, we are only a page away."

5. Some organizations will ask questions about psychological/psychiatric evaluations and fitness for duty evaluations.
   a. Fitness for duty evaluations have a statutory basis for some employee groups (e.g., police officers), but not most (Stone,

2000). In other cases, there is no legal definition. Often, fitness for duty designations have been used to force employees back to work. Employees have been effective in using the leverage of fitness for duty. In contrast, many companies are moving away from this and from premature movement into the medical management of threat cases.

b. Regarding the role of mental illness, we have found that the employee's conduct/behavior in the workplace leads to the violence problems. Therefore, it is best to view this evaluation as a work-related review of unacceptable conduct and not as a psychological or psychiatric evaluation.

c. If the employer presses the question of a psychological or psychiatric evaluation to assess the workplace behavior, the consultant will point out that a diagnosable mental illness is rarely found to play a role in this type of situation. Specialized interviewing skills, not degrees, are needed to conduct threat assessments. (Example: secret service agents conduct such interviews very well.)

d. If the organization wants some professional help for the employee, the EAP or a similar resource should be utilized. This referral may be available while the threat assessment is proceeding. The threat assessment is not a health service. It is a management service. All information may flow to management and no relation is established between the threat assessment professional and individuals interviewed. The consultant will reinforce with the client this difference between threat assessment and clinical help.

e. This type of workplace behavior, once reported to the organization, needs to be reviewed by someone. If an organization does not have the ability to structure a review process, the threat assessor will develop a process that helps to maintain control and can be the one who asks the difficult questions. This allows the employer some time to develop an understanding of what is happening and to develop a management plan to deal with it.

6. All participants should keep in mind that different corporate cultures are operating and must be understood as early in the intake process as possible, e.g., decision-making structure, cen-

tralization versus decentralization, information seeking versus decision making.

7. Identify if there is a need for a security vendor. If so, the consultant may offer to help arrange what is needed. It may also be useful to discuss issues such a stay away letter (see Appendix B), a dummy voice mail box, or a temporary restraining order.

8. The management team should have the consultants' contact information (pager, mobile, and office numbers), and/or emergency dispatch—555-555-1212. Be sure that twenty-four-hour availability is established and that management is encouraged to use the access as needed.

SECOND STAGE OF CONTACT

When the client is familiar with the consultant's philosophy and method of practice in threat assessment cases, the next step is to conduct a conference call with key individuals who have knowledge about the case or who are responsible for making employment decisions.

During the conference call the consultant should cover the following areas:

1. Details of who is on the conference call:
   - Names of employees on the call
   - Titles of employees on the call
2. Name of individual of concern, if known
3. Name(s) of targets and their relationship to the individual in question
4. Details of the situation that led to the referral to the consultant
5. The individual's history with the company, including date of hire, performance evaluations, disciplinary history, and union representation status
6. Tell the client how the assessment will be conducted and how the results of the assessment will be communicated to the client at the end of the consultation. For example, an International Assessment Services (IAS) procedure is to assist the client to think about the IAS category system of classification. Explain briefly to the client each of the five categories used to communicate the assessed potential for violence in the workplace.

- *Category 1:* Available data suggest high violence potential; behavior suggests immediate arrest, hospitalization, or major organizational response
- *Category 2:* Available data suggest high violence potential; behavior does not suggest immediate arrest, hospitalization, or a major organizational response
- *Category 3:* Available data suggest insufficient evidence for violence potential; sufficient data support perception of knowing infliction of distress upon co-workers/others
- *Category 4:* Available data suggest insufficient evidence for violence potential; sufficient data support perception of unplanned infliction of distress upon co-workers/others
- *Category 5:* Available data suggest insufficient data for violence potential; insufficient data of behavior resulting in distress upon co-workers/others

7. In cases where communications are available for review in a written, audio, or video format, or relayed to the organization by a witness, it is useful to educate and describe the principles by which the communication will be reviewed and integrated into the overall assessment. It is also useful to offer the IAS principles of communication analysis factors in examining communications for the level of threat they represent.

- *Organized versus Disorganized:* Level of organization and coherence present in the communication
- *Fixed:* Degree that individuals have concluded that their employment and personal problems are the fault of the company
- *Focused:* Degree that individuals have concluded that their employment and personal problems are not only the fault of the company, but also that one or more specific employees of the company are responsible for the situation
- *Action Imperative:* The need to take personal action to harm the company and its employees because all other avenues are not going to help to resolve the problem
- *Time Imperative:* The need to take personal action to harm the company now or in the immediate future

After receiving a comprehensive and complete briefing of the details associated with the case and providing the client with a structure and philosophy from which the threat will be assessed, several options can be offered as to how management may choose to proceed. Offering management several options enhances their sense of control as the consultants proceed with the case. The three options are:

- *Option A:* Develop a plan to conduct an initial management interview with the employee in question. This initial management pass is often conducted by a human resource or management professional who is internal to the organization and experienced in interviewing. The consultant may offer to work with management to develop and master a set of questions. Make clear that answers should be recorded carefully. Sometimes it will be appropriate to proceed with an internal management plan. However, if interviewers receive other answers that represent a greater potential for threat, which would include communications or behavior consistent with category one or two, they need to contact the threat assessment team immediately. Management, driven by the need to quickly resolve this issue and get on with business, may ask that a list of questions be sent to them so they can take it from there on their own. The consultant should decline and explain that successful interviewing requires a skill set that goes beyond simply asking the questions. The process of the interview is critical; therefore, the scripting of the question with potential branches and nuances of the process needs to be explained in the oral briefing. This interaction allows the internal interview to be repeated, thus ensuring that it was understood.
- *Option B:* The external consultants or those from the team come on-site to interview the collaterals (individuals who have information relevant to the concern) and the employee in question.
- *Option C:* This is a reactive position in which managers decide that they will wait to see if something else of concern happens. If more unacceptable behavior emerges, they will consider either Option A or Option B. While they wait to see what may happen, they need to consider security enhancements or other measures that protect the safety of the workplace and create an environment where the employee in question's behavior is more closely monitored. If the communication is anonymous, specific strategies to ensure the capture of any and all communication

coming into the company via telephone, Internet, or mail is monitored carefully. Some organizations will have no security plan or resources available. The threat assessment professional can assist in providing access to security resources familiar with risk control work.

Additional issues the threat management team should consider:

1. Is the consultant available twenty-four hours a day/seven days a week?
2. If a request for a direct consultant response is imminent, step off-line and determine what staff resources are available and, when necessary, what security resources are available.
3. Signing of a services agreement is very important. The consultant will attempt to get this signed as early in the process as possible.

Chapter 5

Category Classification

Management typically becomes aware that a threat problem exists when a behavior or series of behaviors or comments filter through the management layers. In the previous chapter we addressed the initial action process to begin asserting control over those factors where possible. As a part of that process, information may flow rapidly into the organization from a variety of both confirmed and unconfirmed sources. Events may move very rapidly and teams need to be prepared for this fast pace. The developing picture of the situation may change from moment to moment.

Ongoing research (Cannon-Bowers and Salas, 1998) has shown the utility of a shared model in such situations. A shared model allows members of a team to rapidly communicate perceived changes in the situation. Members also begin to anticipate the informational needs and likely actions of other team members. Team members also begin to weigh options and predict the likely impact of new information on the roles and actions of one another. This should improve the speed and quality of decisions, especially where rapid response may be necessary.

Such a model needs to organize and classify information the team collects and lead to a series of risk control action choices. In addition, such a model needs to be

- Brief and focused
- Easily learned and understandable
- Consistent with the core disciplines of threat management

Our group has used the Threat Assessment Model (TAM) over the last eight years. The late Chris Hatcher, PhD, was the elemental force behind our breakthrough in thinking in the early 1990s. We came to understand that in organizations, actual acts of violence are the tip of the iceberg in terms of what the organization needs to be prepared to manage. A greater scope of challenge exists in the variety of action by members and/or nonmembers of the organization that create emotional distress, fear, and intimidation in others.

Various versions of this category system have continued to develop within our work group, and the model shared here results from the input of a number of individuals. This is a consulting model that is intended to be descriptive of behavior, not prescriptive of personality. This model, and the interview processes supporting it, are not intended or designed as a psychiatric, psychological, or medical examination. It is not a fitness for duty examination (Stone, 2000). Decisions related to specific individuals are placed within a contextual field of perceived threat situations. It is also not intended as a tool for unlimited employment decisions.

The key to the utility of any model is the ability to focus the organization in responding rapidly to risk control actions that need to be modified. In addition, the model needs to be flexible enough to redirect those risk control efforts, if appropriate, when new information becomes available. Teams have a tendency to "label" a situation and then hold to that label. In threat situations risk control actions need to be reviewed on an ongoing basis to determine whether circumstances have changed and whether a decision that was appropriate an hour ago needs to be reconsidered.

Threat behavior does not fall into two distinct categories of violent or nonviolent, threatening or benign. Rather, such behavior extends across a range of actions, situations, comments, and perceptions. Threat behavior includes that which to the reasonable person, intimidates and harasses, and creates emotional distress, fear, and concern. Threat situations also may create distress in particular individuals because of circumstances unique to them.

As stated earlier, the Threat Assessment Model (TAM) consists of five categories related to situations:

> *Category 1:* Available data suggest high violence potential; behavior suggests immediate arrest, hospitalization, or major organizational response

Category 2: Available data suggest high violence potential; behavior does not suggest immediate arrest, hospitalization, or major organizational response

Category 3: Available data suggest insufficient data for violence potential; sufficient data support perception of knowing infliction of distress on co-workers/others

Category 4: Available information suggests insufficient data for violence potential; sufficient data support perception of unplanned infliction of distress on co-workers/others

Category 5: Available information suggests insufficient data for violence potential; insufficient data of behavior resulting in distress in co-workers/others

Each of these categories is intended to organize the available data in a particular situation to focus on necessary risk control actions. Organizationally the objective is neither to over- or underrespond to specific situations.

CATEGORY 1

Available data suggest high violence potential; behavior suggests immediate arrest, hospitalization, or major organizational response required. In this category, the behavior in the situation is of clear, immediate concern and suggests the following.

1. Behavior has occurred that meets the criminal code for immediate arrest. Such behavior may include discharge of a firearm, shooting(s), physical assault, destruction of property, repeated threats of physical assault, repeated threats of property destruction, stalking, or violation of a restraining order.

Team Risk Control Actions

- Ensure emergency call has been made, if appropriate, and determine who will meet the public emergency resource upon arrival.
- Review with local public law enforcement personnel, prosecutor, threat assessment specialist, and labor counsel the behaviors in question and any statements related to motivation, including secondary gain issues, leading to the situation.

- Review legal options to remove the individual(s) from organization site(s) with minimal risk to others and how to deny access regardless of arrest.
- Assess the extent of risk to employees outside the work site.
- Focus on law enforcement role to general public safety issues as well as their assessment of law violations.

The team needs to evaluate the impact of the situation on the organization. Any decisions on employment status or return to work is made by the organization's leaders related to safety and welfare. The decision to have an individual in the organization's areas of influence is not controlled by the public emergency response decision.

- Take action; begin the more detailed risk assessment and subsequent control actions.
- Access employee assistance resources for threatened or distressed individuals.
- Develop, or notify person(s) responsible for assisting the team in developing, a communication plan.

2. Behavior has occurred that suggests this situation meets the criteria of immediate danger of injury to self or other(s) requiring involuntary psychiatric hospitalization. Such behavior may include statements of intent to harm self or others or acting in a manner likely to bring harm to self or others. Decisions for review of involuntary commitment are made by responding emergency services or licensed professionals. Final decisions are typically made at receiving facilities where the individual may be held for one to three days or more for review of potentially suicidal, assaultive, or homicidal behavior.

Team Risk Control Actions

- Ensure that emergency call has been made, if appropriate, and determine who will meet the public emergency resource upon arrival.
- Review with threat assessment specialist and labor counsel the behaviors in question and any statements related to motivation, including secondary gain issues, leading to the situation.

- With consultants, review the likelihood of psychiatric hospitalization. Criteria and interpretation vary from jurisdiction to jurisdiction.
- Transportation and the decision to transport will likely be accomplished by a public emergency response agency. The organization's staff does not make the final decision. They can only initiate the process.
- Prepare information that can be provided to the individual at the psychiatric receiving facility to aid in his or her understanding of the situation. If possible, dispatch someone to brief the evaluator. The receiving facility staff will make the decision whether to involuntarily hospitalize.
- Take action; begin the more detailed risk assessment and subsequent control actions.
- Assess employee assistance resources for threatened or distressed individuals.
- Develop a communication plan.

3. Information may be received in the form of a threat that is of a clear, immediate concern with short lead-time deadlines. Such information may include bomb threats, e-mail/written threats, and/or biological threats. These may require a major organizational response.

Team Risk Control Actions

- Ensure emergency call has been made and determine who will meet the public emergency resource upon arrival.
- Decide whether individuals need to be removed from specific areas.
- Review the situation with local law enforcement personnel as well as threat assessment specialist and labor counsel, if appropriate.
- Take action; begin the more detailed threat assessment and subsequent risk control actions.
- Access employee assistance resources for threatened or distressed individuals.
- Develop a communication plan.

CATEGORY 2

Available data suggest high violence potential; behavior does not suggest immediate arrest, hospitalization, or major organizational response.

In this category, the threat of violence or behavior that could harm is present; however, it is neither imminent nor rises to the level of perceived major injury. The statements of threat in this category are often accompanied by a qualification that if some event does or does not occur in the future then the action would be triggered. Physical behavior may be such that arrest would not occur by the police unless complaints were filed by the target.

Although the reported situation in this category is of concern to the organization, the physical behavior and/or communications are not likely to trigger immediate arrest or hospitalization by local public resources. Such cases may hinge on reports by single individuals with no supporting information as to the exact nature of the incident.

Team Risk Control Actions

- Determine the need for calling emergency services immediately.
- Promptly review available information with threat assessment group, including consultants.
- Ensure that appropriate controls are in place for safety and control of involved individuals.
- Determine what additional behavior/statements would move the situation to category 1.
- Take action, beginning with the more detailed threat assessment and risk control actions.
- Determine legal options to remove the individual(s) from organization sites with minimal risk to others and how to deny access, if required.
- Assess the extent of risk to employees outside the work site.
- Access employee assistance resources for threatened or distressed individuals.
- Develop a communication plan.

CATEGORY 3

Available data suggest insufficient information for violence potential; sufficient data support perception of knowing infliction of distress on co-workers/others.

In this category, the statements and/or behavior that are perceived as threatening harm are the primary concern. The initial report contains no information that would lead to immediate category 1 or 2 concerns, such as actual physical contact having occurred. In this category the statements/behavior are known or should have been known by the maker to create distress in others. The statements/behavior are often made to create sufficient concern that others will not engage in action that the maker believes would be adverse to him or her. They may also be done to intimidate others into action. Such behavior usually has occurred over a period of time, has a clear goal or direction, and/or has occurred in spite of the individual having been advised of the impact of their statements/behavior on others. A reasonable person would know that creating distress is the outcome of receiving such communications.

Team Risk Control Actions

- Promptly review the available information with team and consultants.
- Determine whether behavior rises to the level of review with local law enforcement.
- Evaluate need for removal from organization site based on potential for escalation and creation of further emotional stress on others.
- Take action; begin the more detailed threat assessment and risk control actions.
- Determine what additional statements/behavior would move the situation to a higher category 1 or 2.

CATEGORY 4

Available information suggests insufficient data for violence potential; sufficient data support perception of unplanned infliction of distress on co-workers/others.

In this category, the statements or behaviors are such that others may reasonably perceive them as threatening and experience distress. A single statement, set of statements, or behavior occurs in a limited time frame. The individual engaging in the behavior is unaware at the time of the event how it will impact others. The individual in question has little or no planned goal of creating distress in others. In this category of situation, individuals are able to recognize the reasonable impact of their actions upon the well-being of others. The individuals are typically without a history of distress-creating behavior and accept that such actions are inappropriate and that the organization cannot permit such actions to continue. When confronted with the behavior, action(s), and/or statements they are able to understand the perception, recognize violations of policy, and endorse organizational policy.

Team Risk Control Actions

- Promptly review the available information with at least a subset of the team.
- Determine that no immediate security or law enforcement reactions are required.
- Take action; begin the more detailed threat assessment and risk control action.
- Determine what additional behaviors/statements would change the category.
- Determine where the potential for creating additional distress or misunderstanding exists while the more detailed review occurs.
- Determine availability of employee assistance resources, if needed.
- Develop a communication plan, if necessary.

CATEGORY 5

Available information suggests insufficient data for violence potential and insufficient data of behavior resulting in distress in co-workers/others.

In this category, the available information reasonably indicates that the reported individual may not have engaged in behavior that created the perceived risk. The motivation/circumstances around the situation are not yet determined. This category may include false allegations or situations where not enough data exist to support one of the other categories at this time.

Team Risk Control Actions

- Promptly determine the need for the team review to either return the case to management or proceed with threat review.
- Take action; begin the more detailed threat assessment and risk control process.
- Review available information with labor counsel and threat assessment professional.
- Evaluate the potential for worsening of situation until review is complete.
- Determine what, if any, action is needed in organization.
- Determine what behaviors on anyone's part would cause the situation to be reclassified to a different category.
- Determine range of actions with labor counsel for reporting individual(s), if false allegation is determined.

CASE EXAMPLES

All names and significant identifiers have been changed. Case examples are actually merges of several cases for this discussion.

Category 1: Arrest

Harry has been an employee for eight years and has average job performance reviews. Harry's work unit has been undergoing reorganization as the installation of a new technology is reducing the need for some positions. Although Harry's position is not directly affected, he has been angry and vocal about the changes. He states that these changes are occurring at the direction of younger, less experienced managers who don't understand the work. He has expressed these opinions in team meetings and to other employees while in the break room.

Recently, Harry has focused his dissatisfaction upon Ramon, a middle-level manager, as being responsible for these changes. One day, after a team meeting during which Ramon announced yet another new procedure, Harry muttered under his breath, got up, and stormed out of the team meeting. About an hour later, Ramon encountered Harry in a hallway. Harry seemed to hesitate initially, then came straight toward Ramon, backed him into a corner, and yelled into Ramon's face: "You can't do this to people. Management just ruins other people's lives and then walks away. You push and push and you just might not make it home tonight!" As Harry walked off, Ramon was shaken. A manager for five years, he has encountered a lot of unhappy people, but nothing like this. Walking back to his office, Ramon thought, "Harry was just blowing off steam." Then he remembered that no one had ever spoken to him like that.

Ramon decided to call the plant manager, Chaz. Chaz called security and human resources, who accessed the threat assessment consultants. They quickly decided as a risk control action that security and management personnel should immediately move to determine where Harry was at the moment. Co-workers indicated that Harry said he was going out to his truck in the parking lot on his break. Management, security, human resources, and the threat assessment consultants quickly agreed that it was important to maintain visual contact with Harry and to talk with him, if feasible. Security was requested to notify the local law enforcement liaison that the company may have a problem with a threatening employee, and that further information will be forthcoming shortly. Security was directed to find Harry and, using a cautious, conservative approach, talk with him if possible. Minutes later, security reported that they found Harry in the parking lot by his truck with the truck door open. Harry was conversational, and they talked with him for a few moments. However, while talking to him, one security officer observed what he believed was a sawed-off shotgun sticking out from under the front seat on the passenger side.

The team discussed the fact that, if the security officer's observation was correct, based on counsel's advice, possession of such a weapon was probably illegal as well as in clear violation of the company weapons policy. Based on Harry's threatening behavior/comments and the potentially illegal weapons report, management called local law enforcement and requested assistance. After being briefed on the situation, the two police officers proceeded to Harry's truck in the parking lot where Harry remained. The officers asked about the weapon. Harry denied that any weapon, legal or illegal, was in the truck. The officers saw the barrel of the weapon protruding from under the seat of the vehicle. They knew that Harry was not being truthful and they took possession of a sawed-off 12-gauge shotgun with two live rounds and the safety off, as well as a 9-mm pistol with a fully loaded clip. In the process of taking custody of the weapon, a struggle occurred and the officers cuffed Harry. As he was taken into custody, Harry yelled: "I'll get you. Don't think this is over. Hear me."

In this case, the following actions occurred:

1. Supervisors were aware of a low-level problem with an employee and attempted to solve the problem, i.e., Harry was unhappy with the changes so supervisors and management attempted to communicate new changes and to be available to listen to employee concerns about ongoing change.

2. One supervisor identified a more significant problem when the employee acted in an inappropriate manner toward him and sought consultation from management in an attempt to solve the problem, i.e., Harry made a threat that potential harm could befall Ramon.

3. Management analyzed the need for a plan of immediate action, and sought consultation from human resources, security, and the threat assessment specialists.

4. Management implemented an immediate action plan of determining Harry's whereabouts and having security attempt, with caution, to talk to him, while concurrently notifying law enforcement that a problem may be present with a potentially violent employee.

5. Management was informed by security that a weapon had been observed in the employee's truck, in what appeared to be a clear violation of company policy and potentially the law.

6. Management did not know definitively that the security professional's observation was correct but determined that they had reasonable cause to act and called for local law enforcement to respond.

7. When law enforcement responded and met Harry beside his truck in the parking lot, Harry lied about the weapon in the car, resisted the officers, and threatened Ramon.

8. Management, with consultation from the risk assessment specialists, directed security and the legal department to establish a line of communication with both law enforcement and the district attorney's office, planned for on-site security enhancement in the event that Harry would make bail, directed security to deny Harry access to the company property, informed selected employees of Harry's arrest and his denial of access to company property, and informed selected employees of their responsibility to inform the company if contacted by Harry.

9. Management suspended Harry pending an internal management review of his behavior at work and its potential violation of the company's safe workplace policy, notified Harry of the situation by delivery (which required signature) to his home, and planned for a similar notification to his attorney if Harry remained in jail.

10. Management provided Ramon, as a threatened employee, with access to the company EAP if he wished, and indicated that management would be working with him to weigh the options of continuing at work versus time off or time away to reduce his risk.

11. Management consulted with the risk assessment specialists in psychology to determine what type of additional behavior by Harry would increase the risk pattern and discussed possible concurrent on-site security enhancement.

12. The team developed a telephone notification tree to be used if Harry was to be released on bail from jail, as well as a determination of the utility of security enhancement procedures to be reconsidered in that event.

13. Management consulted with external labor law counsel to assess the option of a civil restraining order.

Category 1: Hospitalization

Fred has been an employee for seven years and has average to above average performance reviews. He has been away from work for four consecutive days. On the fifth day, Friday, Fred appears at work at 8:00 a.m. He proceeds to his work area, starts his computer, and appears to be shuffling papers. At 8:15 a.m., one of his co-workers approaches and says: "Fred, we thought you might be sick or something, so we called your home each day. All we got was your answering machine. Are you OK?" Fred responds with: "No, leave me alone!" His co-worker quickly obliges. A few minutes later, Fred's supervisor appears: "You know, Fred, it is the company's policy that if you are sick more than three days, you have to bring in a doctor's note. While you were gone, the work has been piling up. I sure could use your help with . . ." Before Fred's supervisor can finish, Fred interrupts with: "Leave me alone now!" Later, another supervisor and the human resources manager approach Fred and are met with an additional, escalating angry comment: "Get away from me now or I'll make you get away!"

Fred draws an eight-foot circle around his area and will not allow anyone inside it. Everyone is now concerned, but unclear as to what to do. Security is notified, and assumes a low-profile position nearby. Other employees are directed around Fred's work area while the problem is being discussed. As requested by the human resources manager, two risk assessment specialists in psychology arrive at 9:00 a.m. Sitting on the very edge of the eight-foot circle and backed up by security and with local law enforcement on standby, they try to communicate with Fred: "Fred where have you been for these past few days? What has been happening?" Slowly, Fred begins to respond: "I've been at home waiting for my supervisors to come and get me." A few minutes later, Fred continues: "My supervisors came to my home. They came up to the window and had long glistening knives. They were looking for me. I was scared so I closed the blinds quickly and went into the closet and hid." Fred then turns and looks directly at the risk assessment specialists in psychology for the first time, saying: "You're taking me to the locked ward, aren't you?"

The employee has lost contact with reality. At present, he perceives his supervisors to be a threat to his life and is attempting to avoid them. However, both Fred and the risk assessment specialists in psychology know that Fred may not be able to continue his high level of defensive vigilance. Unpredictably, Fred may react to his supervisors or co-workers in a preemptive manner to hurt them before they hurt him. While continuing to talk with Fred, the risk assessment specialists in psychology and security notify law enforcement of the nature of the situation. Two law enforcement officers respond to the company site, are briefed on-site, and are introduced to Fred, who repeats his concerns.

The officers proceed to transport Fred to the local locked psychiatric facility. Knowing that the transporting officers will not have much time to spend at the hospital to brief the psychiatric admitting staff, the two risk assessment specialists in psychology along with company security follow the patrol car to the hospital in other vehicles. Fred is subsequently admitted, with a request by law enforcement that they be notified of any impending release.

In this example, the following steps occurred:

1. A supervisor identified a problem with an employee and attempted to solve the problem, i.e., Fred was angry and minimally communicative, so the supervisor tried to talk to him.

2. Supervisors and managers identified a more significant problem with an employee and attempted to solve the problem, i.e., Fred was angry and nonresponsive to the communication efforts of his supervisor and team members, so management tried to talk to him.

3. Supervisors and managers identified a very significant problem with an employee, i.e., Fred was angry, nonresponsive to communication efforts, and had begun acting in a manner that others perceived as threatening, so management sought consultation/assistance from security and risk assessment specialists.

4. Management focused on the safety of Fred and co-workers by bringing security to the scene, keeping other employees away from Fred's work area, notifying law enforcement of a potential problem, and requesting immediate on-site response from risk assessment specialists.

5. At the direction of management, risk assessment specialists determined that Fred's behavior was indicative of a category 1.

Law enforcement personnel were requested to respond to the site and were appropriately briefed on the situation. Law enforcement personnel were introduced to Fred so that Fred could restate his fears and concerns. Fred was deemed a probable candidate for involuntary psychiatric hospitalization and was transported by law enforcement.

6. At the direction of management, security and the risk assessment specialists proceeded to the hospital where Fred was to be admitted, so that the admitting health care professional would be fully aware of Fred's behavior in the workplace. Law enforcement requested that they be notified if Fred was to be released in case they were required to respond again to a situation involving Fred in their community.

7. Management, with consultation from the risk assessment specialists, planned for on-site security enhancement in the event that Fred should be released from the hospital, directed security to deny Fred access to the company property, informed selected employees of Fred's behavior at work and the denial of his access to company property, and informed selected employees of their responsibility to inform the company if contacted by Fred.

8. Management placed Fred on paid leave, pending an internal management review of his behavior at work and its potential violation of the company's safe workplace policy, and notified Fred of these events by a method requiring signature at his home and at the hospital.

9. Management provided Fred's supervisors and other employees who had contact with Fred that day access to the company employee assistance program, if desired.

10. Management consulted with the risk assessment specialists to determine what type of additional behavior by Fred would increase the risk pattern and to discuss possible concurrent on-site security enhancement.

11. Management consulted with external labor law counsel to assess the option of a civil restraining order, as well as to discuss other employment considerations.

In these two examples, immediate removal of the individual from the workplace, determining the utility of different security enhancement procedures, and the overall prevention of any violent episodes are the primary concerns of the threat response and managerial team.

Category 2

Steve has been an employee for twenty-two years, and had good performance reviews until the last eighteen months when his performance declined first to below average and occasionally to unacceptable. Four months ago, Steve injured his foot while assisting in moving a large item of company equipment; he filed a worker's compensation claim and was away from work for one month. During that time, Steve called his supervisor and various co-workers, complaining that the injury was very painful and he was receiving little assistance from his doctors. When Steve returned to work, he seemed to do all right for two months. Then his attendance and tardiness became problematic. When confronted about this, Steve replied that his foot still hurt.

Meanwhile, Steve's employer was the object of a takeover by a much larger competitor. After the announcement of the takeover, months dragged on with the employees wondering how many jobs would be lost when the two companies completed the merger. Steve was particularly angry. His blue-collar technical skill had earned a good wage for him and his family. However, his technical skill was industry specific. If he lost this job, there would be few alternative job opportunities, and none that would begin to approach his current wages and benefits. Not only was the company going to betray his loyalty, but they also were leaving him with a painful injury that would restrict his ability to get another job. Steve became more and more bitter.

He wrote a long letter to the worker's compensation agency in which he talked about how he had put all his adult working life in this company and now the company was just going to leave him with this endless pain. Early one morning, Wayne, a co-worker, observed Steve standing in the distance and went over to see if he was all right. Steve said to Wayne, "You know, I finally figured it all out. The company doesn't really care about anybody. If you get hurt, they just find a way to throw you away. But they're wrong if they think that's going to work with me. I'm not going to live out my life in this kind of pain with no money. I bought a .357 last week and if they lay me off, I am going down that glass hall. First, I'll get Jenkins, and then I'll go over to Overman's and finish him, too. I know exactly how I'll do it. I've been dreaming about it. I really have nothing to lose. But you don't have to worry. I'm not angry with you. When I call you, don't come to work that day."

Wayne had been to the company's workplace violence prevention training session this year, and the trainer said to pay attention to such threats and report them to management. Wayne had known Steve a long time. Steve always lived up to what he said, and this bothered Wayne much more than the training instructions. After agonizing about reporting the incident and talking to a close friend, Wayne could not bear the responsibility of keeping quiet about Steve's statements. Besides, any harmful act Steve did to others would make the situation worse for Steve. When management listened to Wayne's report, they became concerned as well, for they knew Steve as a man of his word. As management assembled the workplace violence team with representatives from human resources, security, legal, and risk assessment specialists in psychology, Steve was reported by his supervisors to be back at work, doing his job.

Management consulted Steve's supervisor, who reluctantly acknowledged that he had overheard Steve make a similar comment to someone early this morning, and that Steve had told him last week about buying the gun. Neither Wayne nor Steve's supervisor knew that management had already tagged Steve's position for elimination next week. The management team decided that there was reasonable cause to further investigate Wayne's report of Steve's behavior, and should do so cautiously. An initial management discussion with Steve was discussed as an option, as was proceeding directly to an on-site interview by the risk assessment specialists in psychology with security backup and law enforcement on standby. The management team decided that the latter approach would minimize risk and maximize the potential to obtain the most useful information.

The risk assessment specialists' interview of Steve covered his work history with the company, his injury, and his statements earlier in the day. During the interview, Steve acknowledged his prior statements to Wayne. He confirmed that he had nothing to lose but his pain if the company laid him off and he needed to take action. Wayne also confirmed that he had bought a gun and concluded with: "As long as I have my job and my paycheck, everything is OK." The risk assessment specialists found the situation to be within category 2. The management team reviewed this information, examined their options at this point, and decided that Steve should be removed from the workplace by administrative leave. His access to the company site would be denied, relevant managers and supervisors would be notified, and security enhanced until further information was available. The option of a restraining order would be examined with external labor law counsel.

The management team also alerted relevant individuals to the importance of vigilance with regard to further phone calls, visits, drive-bys of the company, drive-bys of residences of company employees, or letters that might indicate a continuing escalation of Steve's need to let the company know that he was out there and angry, and possibly indicate an escalation to violence.

In this example, the following steps occurred:

1. Steve was a long-term employee, had a good job performance history that declined, and had sustained an on-the-job foot injury, which continued to be painful. The employee was appropriately cleared medically for return to work, and had made allowances for his reentry into his normal work routine. The employee was cleared for work a second time when he continued to complain of foot pain at work.

2. Steve told a co-worker that he would take action to kill others if he lost his job. This threat contained references to having re-

cently purchased a handgun, to not having anything to lose if he killed others, and to the unremitting pain in his foot for which he blamed the company.

3. The co-worker informed management, and management then confirmed concern about Steve's behavior by talking directly with Steve's supervisor.

4. Management promptly convened the workplace violence prevention team, examined options in proceeding to interview Steve, and determined that risk would be minimized and information gain would be maximized by having this interview conducted by risk assessment specialists.

5. During the risk assessment specialists' interview, Steve acknowledged his prior threatening statements to co-workers and his own moral belief that such action was justified if he was terminated or laid off. Such behavior is characteristic of category 2.

6. Management placed Steve on administrative leave, pending an internal management review of his behavior at work and its potential violation of the company workplace policy, notified Steve of these events, had security escort Steve to his car, and notified Steve again of his administrative suspension by a method requiring signature at his home.

7. Management, in consultation with the risk assessment specialists, provided Steve's supervisors and other relevant employees with information on the EAP.

8. Management consulted with the risk assessment specialists in psychology to determine what type of additional behavior by Steve would increase the risk pattern and to discuss possible concurrent on-site security enhancement.

9. Management consulted with external labor law counsel to assess the option of a civil restraining order, as well as other employment considerations.

Category 3

Mary has been an employee for five years and has had below average job performance reviews. She has been counseled about her deficiencies twice, managing in response to marginally correct the problems. Mary became aware that the two co-workers in her pay grade on her team were getting merit increases. She was not. Mary was angry about this and convinced herself that she was being treated unfairly. She thought, "It is only a 5 percent merit increase, and workers at my pay grade need every nickel to survive." Mary decided that she would get a notebook, watch her co-workers, and document every failing in order to show that they were no better than she was.

Mary obtained the notebook, and was even observed once or twice by her supervisor, Jerri, writing hurriedly into the notebook. Jerri did not know what was occurring. There was enough to worry about in the office as it was. Several weeks later, Mary's two co-workers came to Jerri, their supervisor, to complain that Mary seemed to be watching them, although they did not know why. Jerri made a mental note to talk to Mary at the next opportunity. The next afternoon just before quitting time, Mary approached Jerri and said: "The other two employees in this office have been getting their merit increases and I'm not. If you don't give me my merit increase, I guess I'll have to get a gun to make this happen." When management listened to Jerri's report, they were concerned and surprised. Mary was known to be opinionated, even intimidating, but she had never threatened anyone.

Management assembled the workplace violence team with representatives from human resources, security, legal, and (by telephone) risk assessment specialists. The team decided that there was reasonable cause to further investigate Jerri's report of Mary's behavior and that they should do so cautiously. An initial management discussion with Mary was an option, as was proceeding directly to an on-site interview by the risk assessment specialists in psychology with security backup and law enforcement standby. The management team decided that the latter approach would minimize risk and maximize the potential to obtain the most useful information. Mary had left for the day, so the interview was scheduled for first thing the next morning.

The risk assessment specialists' interview of Mary covered her work history with the company, her failure to get the merit increases, and her statements earlier in the day. During the interview, Mary acknowledged her prior statements to Jerri. She stated, "If it was necessary for other people to be distressed and fearful in order to get the merit increase that I deserved, then that was just fine." Mary went on to explain that not receiving her merit increase had caused her a lot of distress. She felt entitled to "pass some of that distress around." Further, Mary related that she had told her older son what she had said the other day at work. He had told her such statements would be upsetting to others. Mary said that she had told her son in response that she "deserved the 5 percent no matter whether others got upset or not."

The risk assessment specialists found the situation to be within category 3. The management team took this information and examined their options. They decided that Mary should be removed from the workplace by administrative leave and her access to the company site denied. Relevant managers and supervisors would be notified. Management, in conjunction with human resources, would then determine whether Mary's behavior sufficiently violated company policy to warrant termination, or to warrant suspension with an associated return to work agreement that clearly identifies behavior standards.

In this example, the following steps occurred:

1. Mary was an average to below average employee who had not demonstrated work performance sufficient to warrant merit increases.
2. Mary's note-taking behavior while watching her co-workers was unusual, but not sufficiently so as to raise attention with her supervisor until other events occurred.
3. The supervisor promptly reported the statement and the associated co-worker behavior reports to management.
4. Management promptly convened the workplace violence prevention team, examined options for interviewing Mary, and determined that risk would be minimized and information gain would be maximized by having this interview conducted by external risk assessment specialists.
5. During the risk assessment specialists' interview, Mary acknowledged her statements, acknowledged her statements as intending to cause distress and concern in others for her own self gain, and acknowledged that family members had also warned her that such behavior would be distressing to others. Such situations are characteristic of category 3.
6. The management team took this information, administratively suspended Mary, and proceeded to determine whether Mary's behavior sufficiently violated company policy to warrant termination, or to warrant suspension with an associated return to work agreement that would clearly identify workplace behavior standards and require that Mary adhere to those standards.

Category 4

Huang is twenty-eight years old, has been an employee for four years, and has above average job performance reviews. It is Halloween, and many of the employees are talking about their kids' costumes this year, which parent will be taking the kids around this time, and whether it will rain heavily like it did last year.

At lunchtime, fifteen employees are on their break. The break room has one point of entry and exit, no windows, and overhead fluorescent lighting. At about 12:30 p.m., Huang enters the break room, wearing a short windbreaker-type jacket. With his elbow, Huang nudges off the overhead light switch, placing the room in semidarkness. With his right hand, Huang takes out a plastic replica .45 caliber handgun, which is actually a water pistol. With his left hand, Huang quickly removes a lighted jack-o'-lantern from his windbreaker, and yells: "Let's all play post office!" As some employees dive under the metal tables in the break room and others cry out in distress, Huang shoots water from the water pistol at some of his co-workers on the floor.

Huang's co-workers are not amused. In fact, some are angry enough that Huang decides to leave after returning to his work area. Huang's supervisor and manager quickly interview him. He attempts to explain that it was a Halloween joke that seemed like it would be funny at the time. Huang has now realized that his idea of a joke has been very distressing to his co-workers. When one manager likens the joke to crying out "Fire!" in a crowded movie theater, Huang agrees, asks if he can apologize to everyone, and states that he will do anything to keep his job.

In this example, the following steps occurred:

1. Huang was an above average employee with no demonstrated job performance problems or interpersonal problems with co-workers.
2. Huang's behavior, although apparently meant as a practical joke, distressed his co-workers, some of whom remained distressed while others became very angry. The impact upon production at the site that afternoon was significant.
3. Huang belatedly recognized his error, wanted to apologize to his co-workers, and was willing to do whatever was necessary to save his job.
4. Management consulted with the risk assessment specialists, provided performance counseling with Huang, convened a work team meeting, and developed a return to work agreement that specified acceptable behavior and the consequences of further incidents.

Category 5

Francine had been an employee for six years, with average job performance reflecting some problems of attendance and tardiness. She got along well with co-workers and had been a good team member. Francine worked in production from 3:00 p.m. to 11:00 p.m. weekdays.

One day, at about 6:00 p.m., Francine approached Tanya, one of her co-workers, and asked: "Will you go on my smoking break with me?" Tanya replied that she doesn't smoke, and wondered why Francine was asking her to do this. Tanya decided that maybe Francine just wanted the company, so the two went outside to the employees' smoking area adjoining the parking lot. Francine lit her cigarette and went to the urn filled with sand. Pausing a moment, Francine grabbed Tanya's arm, pointed to a car parked in the parking lot, and cried: "There's a bomb on the dash of my car!" The car was forty feet away, and Tanya could not really see anything.

Security was called. The officer could indeed see a device on the dash. The device had two red cylinders with wires attached to a battery. The threat management team contacted law enforcement who responded. The threat management team also directed that evacuation procedures be implemented. When the bomb squad used their portable X ray to investigate, the device was discovered to be a fake. The threat management team and law enforcement faced overlapping problems. Each had their own interests in trying to determine who had placed the fake device in the car.

Security, risk assessment specialists, labor law counsel, management, and law enforcement all needed to work together in this review. As the coordinated review continued, it appeared that situations similar to this had happened before in Francine's life. Whenever her marital life was not going well, situations that placed Francine in the role of victim occurred.

In prior situations, no one was ever identified as Francine's persecutor. The threat team recommended a cautious position, as no evidence existed that Francine had created this situation. Weeks later, Francine confessed to the event, and she asked the threat management team to be allowed to resign.

In this example, the following steps occurred:

1. Francine was an average employee with no significant performance or interpersonal problems with co-workers.
2. The threat management team followed the designated response plan, notified law enforcement, and conducted an appropriate evacuation of the potentially affected work area.
3. The threat management team was informed by law enforcement that the device was a fake.
4. The threat management team conducted a cooperative joint review of the event, in coordination with legal counsel, security, and the risk assessment specialists.
5. The threat management team was informed by Francine of her involvement in the incident. She requested to resign, and did so.

In summary, this approach offers a multicategory system for understanding and organizing the flow of individual case information to the threat management team. Organizations will need to determine the usefulness of the Threat Assessment Model (TAM) for their unique situations. Further, the examples cited in this chapter are for illustration purposes only. Each case of violence or threat is unique and warrants consultation with labor law counsel and risk assessment specialists. The model does illustrate the importance of a thoughtful system in order to effectively deal with the potentially violent/threatening situation, the information flow, and risk control actions.

Chapter 6

Threat Management: Interviewing Strategies

INTRODUCTION

Threat management continues to be an evolving specialty. Violence in the workplace, the home, and the community has become an increasing concern for the public and for lawmakers. In the past several years, significant legislative changes have occurred in regard to communicated threats, stalking, harassment, and domestic and workplace violence. Although many labels are attached to violence, depending on the context it occurs in, "threats" is the most appropriate label for use here.

As a result of these legislative and organizational policy changes, significant demand has been placed on human resource professionals and corporate security regarding investigation and management of these organizational challenges. In the past, laws and policies regarding terroristic threats and stalking were considered weak and required a very high threshold before employers or law enforcement responded. In most cases, employers would transfer or ignore employees. The police were required to wait for an overt action, such as an assault, abduction, or attempted homicide, before conducting a formal investigation.

Today the requirement for employers and law enforcement to respond to communicated threats has been modified. Stalking laws are in effect for all fifty states and the District of Columbia. Employees and employers are less tolerant and more willing to report communicated threats and violent behavior in the workplace. The tolerance threshold has been lowered as a result of significant media attention and workplace violence awareness training. The number of reports of communicated threats, acts of harassment, and intimidation to human resource and management in the workplace has increased.

More recently, there has been considerable effort by employers to train senior managers, supervisors, and employees on this issue in the workplace. Police officers are trained to evaluate threats and develop strategies to intervene earlier in cases where the risk is considered to be significant (Fein and Vossekuil, 1998).

Threat assessment and management has become more visible in corporate America. Corporations have developed new strategies and tactics to deal with organizational crises. This is evidenced by the increased number of policy statements that reflect a change in company tolerance for inappropriate and threatening behavior in the workplace. Many companies have formed incident/threat management teams to assess threatening communications and behavior in the workplace and to direct risk control actions.

A violent act is the culmination of multiple factors that reflect problems, conflicts, disputes, and failures (Fein and Vossekuil, 1998). It is not a random act but develops over time, like a storm (Monahan and Steadman, 1996). It occurs when there is an interaction of three factors: the characteristics of the individual, an event or situation that triggers a reaction, and conditions that allow violent/threatening actions. Individuals who engage in a violent act often feel helpless and desperate. In their minds, the violent act is perceived as a viable solution to a situation in which there are no perceived solutions. For some individuals, threatening statements, actions, and/or behaviors are a primary form of communication. Unchecked by management and policy, they expand their sphere of influence creating a zone of fear, intimidation, and emotional distress that inhibits the ability of the organization to perform its mission. This sphere of fear degrades the lives of employees, customers, supervisors, and other stakeholders by distorting the work processes and creating a potential hostile environment.

INTERVIEW AND INVESTIGATION

The collection of data for threat assessment and risk control actions occurs primarily through interviews, reviews, and investigations.

Rather than challenge themselves with long-term predictions of violence, managers of organizations can focus on risk and reducing the factors related to it. Theoretical positions support the assessment of an individual's communications and actions along a continuum. This assessment is based on patterned behavior that has been observed in the past.

The following themes are offered to provide clues as to how an individual may move along a continuum from idea to action.

1. Individual suffers trauma, which creates extreme tension, anxiety, and/or desperation. A single major event (layoff or termination, actual or perceived), or a series of minor events wherein the last incident represents the proverbial straw that broke the camel's back, life events (divorce, end of romance, loss of significant other).
2. Individual perceives that problems are essentially unsolvable, usually after a series of attempts.
3. The individual projects all responsibility onto the situation, blaming an organization, an individual, or both.
4. Individual's frame of reference becomes increasingly self-centered. What is happening, e.g., layoff, court action, etc., is viewed as being directed specifically at him or her.
5. As a result of feeling persecuted the individual shifts from a position of self-protection to self-preservation. Self-preservation becomes the primary objective.
6. Threats/violent act perceived as the open path. The individual may communicate or express violent ideations. The degree to which this type of communication has worked in the past impacts the speed with which it becomes the primary option over others.

   This may be reflected in a statement of intimidation, designed to inflict emotional distress, or the individual may be more descriptive with plans regarding an action or time imperative. The individual may not necessarily verbally communicate anything but may engage in behavior such as rehearsing with weapons or following or pursuing potential targets.
7. Finally, the individual commits to violence as the solution and the violent act is attempted or committed.

As a result of growing attention, organizations are required to conduct reviews of behavior(s) of concern. This type of review requires different strategies compared to other kinds of interviewing in the organization setting (hiring, operational, customer focused). These situations require thoughtful assessment, review, and special attention to interviewing strategies. Unlike crimes or situations where a variety of investigative resources in crime scene analysis create the foundation for further review, organizational settings are limited to reports of behavior that may or may not have been witnessed by more than one person and physical data, if any, is often very limited. The organization's decisions about communicated threats are built around the thoughtful analysis of the data and the results of the interview with the reporting individual(s), collateral witnesses, and the individual in question, all or some of whom may be employees.

The main purpose of an interview is to assess and establish whether the situation is one of making threat(s) or one of actually posing a threat and whether the alleged perpetrator is moving forward in this process. The interviewers identify what appear to be the critical risk factors that need to be and/or can be influenced. The threat assessment team processes where the subject is from creating emotional distress to actually carrying out threatened/violent behavior and what, if anything, needs to be done to interrupt the forward motion. The identification of what may increase or decrease the likelihood that the subject will choose to move farther along the path becomes the focus. These critical variables, once assessed, become the basis for risk control actions, which must be monitored once in place. Often the clues to these questions lie with the recipient, collateral individual(s), or the individual reported to have engaged in the behaviors. Gathering this critical intelligence is essential to formulating the appropriate response to the threat situation.

Some of these situations can continue for years before coming to the attention of the threat management team. Individuals may have trouble connecting and recalling all the incidents, many details, or the proper time sequences. The interviewer must query individuals about all other possible related incidents such as vandalism/property damage (slashed tires), other situations they have witnessed, surveillance, unwanted/unsolicited gifts/flowers, telephone calls/voice mails, letters, notes, and tapes (audio/video). Any changes in the tone, content, or the context of the communication needs to be noted. Any noticeable persons lingering in the neighborhood home or the organization, any burglaries (home or office), or any stolen property should be described in detail.

Noting the times when the subject attempts communication with the individual can be of great importance. For example, a case was referred for a risk assessment, in which numerous harassing/threatening messages were left on an individual's voice mail at work. The suspected perpetrator was a former co-worker and knew that the individual worked between 8:00 a.m. and 4:00 p.m., Monday through Friday. The subject always called between 11:30 p.m. and 1:30 a.m., essentially ensuring that the individual would not be there to answer his calls. Such patterns play an intricate part in addressing risk assessments.

Target Interviews

Risk assessment is an ongoing process that changes with time based on the perception and subsequent actions of the individuals involved. The goal of interviewing recipients and collaterals is to gather enough relevant information to formulate and assess the current risk and to identify possible risk control options.

In any review there are guidelines in regard to sensitivity and advocacy. Reviewers have become increasingly attuned to victims. Ongoing contact between the recipient and human resource personnel is important so that any additional information is more likely to be passed to the team. It also allows human resources to detail questions of how, when, and under what circumstances. This information is especially critical in violent crimes. Ongoing contact is essential in stalking cases or domestic violence cases in which emotions run very high and situations can escalate very rapidly. Often, when a stalking case comes to the attention of the organization, the target has attempted to handle the situation alone and inadvertently often aggravates an already tenuous situation. Targets often become emotionally exhausted and may fear the loss of position in the organization.

Communicated threats are cases that are generally subjective in nature and, therefore, the evidence and the important data may rest only with the recipient. The recipient may provide dimensions associated with the threat to include the context in which the communication was made as well as the history of previous communications and behaviors. The recipient will also be able to identify critical collateral sources who should be interviewed.

Integration of Victim and Collateral Interviews

The results of interviews with the victim and collateral sources help to create an information grid surrounding the communicated threat under investigation. This often will include specifics associated with the communication or behavior, such as time, place, context, specific statements, and behaviors. The collateral information lends itself to a timeline of events that reflect a series of observations of the subject's communications and behaviors. The information grid becomes a data log from which the interview of the subject will be scripted. It will also be a source of data to evaluate the individual in question's version of events. The grid allows the interviewer an opportunity to assess the subject's story and how it may change over time. The grid also allows for the identification and evaluation of inconsistencies, contradictions, and opportunities for corroboration, the question being: Does the information/evidence support the allegations? Contradictions or inconsistencies need to be addressed with the recipient, collaterals, and/or the individual in question.

The interviewer also assesses whether the individual's demeanor and presence fits with the behaviors reported. For example, a recipient reported alarmingly escalating and high-risk behavior on the part of a subject. The interviewer noticed that based on the interview behavior of the alleged perpetrator and collateral interviews, his nonaggressive and almost shy behavior was out of character with the behaviors and actions attributed to him by the recipient. Upon further interview with the recipient, she admitted that she had made a false stalking allegation against the subject for a perceived wrong committed several years ago that she did not feel had been addressed adequately.

The interviewer needs to use an information grid to query the individuals in question regarding the outcomes of their previous behavior. This gives valuable information as to how the subjects perceive the current situation and how much insight they may have on the consequences of past behavior.

Interview Questions

It is important to interview the recipient to fully gain the subtleties of the situation and to provide a high level of detail to the information grid. The relationship between the target and the subject is relevant when conducting an interview with a recipient. The interviewer should specifically explore whether the nature of the relationship is one of the following: spouse, significant other, domestic partner, family member, peer, colleague, co-worker, acquaintance, or no known relationship.

The degree to which a relationship is intimate has proven to have some predictability in regard to the potential for violence in stalking and workplace cases (Meloy, 2000). In the organization, the interviewer needs to obtain the limited information necessary to make a threat assessment and determine risk control issues. In domestic violence cases it is essential to establish whether a pattern exists of behavior and communication that suggest escalation toward violence. The interviewer needs to be able to set boundaries and, at times, refocus individuals on relevant issues related to the situation.

Exploring the nature of the relationship provides a foundation to assess the context of the communication and to identify possible collateral witnesses. It also allows the reviewer to obtain a better understanding of the pattern of behaviors that may create risk. For example, a female co-worker became obsessed and began harassing a male co-worker. She wrote him letters and developed an infatuation. The male co-worker was confused as to why she selected him as her target since he had never encouraged her nor had he barely spoken to her. Over time she began to harass him in and out of the office. When questioned about why she was so interested in him, she promptly stated that he was obviously in love with her since he continually made overt attempts to be around her in the office by walking past her cubicle each day and saying, "Hello." She gave no other reason for her perception that he was in love with her.

In another case, a female reported that she overheard a male talking about what she believed was a shooting scenario. She reported that she was on his sales team. She never had dinner with him alone, only as part of a group with other salespeople. She reported there had been an ongoing conflict over the sales territory. About two weeks prior she reported to the manager that the male co-worker had not called on a shared client in a month. This led to a review, which was still ongoing, as to whether he had falsified visit and expense reports. She reported that at least two other co-workers told her that the male blamed her for the review.

CHRONOLOGICAL DOCUMENTATION

Documentation is an important task in any organizational-related review. In a review associated with threats or violence, documentation may be the only viable record to justify the organization's decision. Documentation is critical from several standpoints.

First, documenting the specific words and language used by an individual in an alleged threatening communication or behavior will become critical in defining and assessing the threat level. Is it a harassing communication or a terrorist threat? These are subjective crimes and the exact nature of words, phrases, and descriptions of behaviors, such as mannerisms and gestures, can often make a difference in appropriate decision making. Whether implied or direct threats are used, what needs to be documented is the patterned behavior over time, if such exists.

Second, the assessment of the situation's threat level is dependent upon the close examination of words and behavior within specific contextual situations. Is the communicated threat a direct threat or an implied threat? Is it vague or conditional? The degree of organization, fixation, focus, action, and time imperative are core variables in assessing where a situation falls on the range of immediate physical threat to creating emotional distress. Such issues are then critical in determining the necessary risk control actions.

Documenting the communications from the victim allows the reviewer to subsequently create an information grid for additional collateral interviews. The grid will offer a comparison to what others may have heard or what others report had been communicated in the past. This process assists in substantiating previous statements. In cases of false allegations, the information grid provides the reviewer insight into the degree of distortion or exaggeration, as well as where corroboration is lacking.

Finally, the documentation of the specific communications and behaviors will later become an integral part of the interview grid to be used when interviewing the individual in question. It will be critical, as will be examined later, to present the subject with exactly what is reported and to assess the verbal and nonverbal response to that presentation. Chronologically documenting the length and frequency of the contact between the recipient and the individual in question as well as validating specific incidents is crucial.

In documenting these incidents it is important to address a series of questions: What was the context of the communication? Was a particular theme repeated? Is there a discernible pattern around anniversaries or other times of the year? We consulted on several investigations where the communication activity appeared to increase around an anniversary of a termination, birthday, or some other major life event. In evaluating targeted violence, the interviewer should check with others in the organization for significant event timing of any incidents reported by the subject.

Checking on previous incidents can determine credibility of the target's reports. In one case a female target reported that she was being stalked and harassed by a male co-worker. The description of the incidents provided by the victim indicated a high risk for potential violence. However, based on a detailed interview of the target and a check of the co-worker's status, it soon became apparent that the co-worker could not have committed many of the alleged acts. On a number of situations that she reported, he was out of the city on business trips.

ESTABLISHING THE PATTERN

Establishing a pattern and a history is important. Equally important is to establish a pattern of outcomes following the incidents of threatening communications, behavior, or targeted violence of both the target and the alleged perpetrator. It is important to get a detailed and specific description of the individual's behavior and communications as well as the context in which they occurred. The review needs to be detailed and capture the essence of the behavior that accompanied the communication. The use of behavior to emphasize aspects of the words used in the communication determines nuances of interpretation.

It is necessary to document the actions evidencing emotional distress. Examples of operationalizing emotional distress are telling someone else, changing the door lock, installing an alarm, and parking in a different space. The impact of a threatening or harassing communication or action upon a recipient is included in the laws associated with terrorist threats, stalking, and domestic violence. The degree to which a victim demonstrates emotional distress as a result of the communication or action needs to be established. In an organizational setting the behavior may represent a violation of policy without the necessity of documented fear. Many states require the target to demonstrate a reasonable fear of bodily harm for law enforcement action; however, the failure to meet a law enforcement standard does not prevent the organization from responding. Documenting this operationalization of fear and emotional distress includes noting the changes that targets make in their daily routine and whether they have sought assistance or used outside services such as a shelter, counseling, social service agencies, or private security.

Did the recipient partially relent to the individual's demands? Were the police called? Was anyone arrested? Did nothing happen at all? Although history of violence has been noted as a significant indicator, the outcome of a communicated threat is important from the standpoint of whether the individual is responsive to limits, such as law enforcement. On the other hand, successful use of threats or physical intimidation will lead the individual to feel reinforced and empowered by successfully gaining control over the situation.

In evaluating the behavior or communication, the reviewer needs to consider asking about behavior indicative of substance use. In an organizational setting where both are employees, the reviewer needs to work closely with labor counsel to determine the extent of questioning in this area.

The reviewer should also explore any patterned behavior that reflects a loss of touch with reality where the subject becomes suspicious. Questioning should specifically identify the beliefs that have been expressed to the target. The reviewer should identify, if appropriate in the target's presence, any history of previous danger to self communications, plans, or self-inflicted injury. In what context did this occur and what was the outcome? Did the recipient succumb to the individual's demands when confronted in the past with a suicidal communication? Were threats of harm communicated to others in the organization who the individual may think are interfering or impeding some resolution between the individual and the recipient?

COLLATERAL INTERVIEWS

Collateral interviews are conducted with individuals who may possess relevant information. In organization settings, one is typically limited to other employees and supervisors. No contacts outside the organization should occur without the review of labor counsel, and then only when these have been determined as absolutely critical. The large majority of internal reviews can be done without involving outsiders. Questions of privacy and control are major concerns when nonemployees are involved. An employee can be directed not to discuss the interview and reminded of potential employment consequences for failing to maintain the security of the interview. No such control exists with outsiders. After contact, they may immediately pass information along and heighten the risk of the situation. As well, the review process should involve the minimum number of co-employees as possible to make an assessment of threat and risk. The collateral interviews consist of interviewing co-workers or individuals who may have witnessed a communicated threat or threatening behavior, or with someone familiar with the subject and/or the recipient of the investigation. Again, issues of privacy require that consideration be given to involving the minimum number of individuals necessary.

The necessity for collateral interviews is to gain a broad sample of the subject's past and current behaviors, actions, and statements as well as to verify those already reported to adequately assess the situation. This provides at least some information and assists the team in developing initial strategies to manage the individual in question. A first pass by human resources or security can identify the range of knowledge of individuals. In many cases, threat assessment professionals will want to reinterview these individuals because they are looking for very specific information to resolve discrepancies and for subtle responses.

Collateral Relationship

Depending on the collateral's relationship with the involved individual, gestures and other nonverbal clues may not only be interpreted in a different manner from statements, but may be more revealing. For example, on consultations in several workplace violence investigations, we found that it is common for the subject to make more revealing statements to certain individuals whom they perceive they can trust, or to those whom they wanted to warn before they committed their violent acts. Also, we found that, following a violent act, many collateral witnesses stated that the subject often talked about his actions, but that they either did not believe that the individual was serious or they did not believe that the individual had the ability to carry out any acts of violence. Believing as they did, they failed to report the individual's behaviors/statements to anyone at the time. Once they are aware that something has happened or that a review is occurring, this information is revealed.

Collateral Interview Strategies

These recommendations for interviewing techniques tend to elicit useful information. This information will be used to make an assessment in regard to the risk in the situation, to assist in formulating an interview grid for the alleged perpetrator, and to consider various risk control options and their likelihood of success.

1. Describe the relationship between the victim and the subject, the collateral and the victim, and/or the collateral and the subject.
2. Describe the length and the frequency of contact between all parties. Has this type of communication occurred previously? If so, what was the context and is there a repetition of a theme from the subject?
3. Are there any identifiable risk factors? Has there been a threat of harm to others connected with the victim, to include family or friends? Have the police been called on other occasions? Are they aware of any previous incidents of violence in any other relationship either with the victim or the subject? What may have been the outcomes of the previous incidents?

4. Document each incident with emphasis on the content, tone, and context of the communication/behavior. Describe the manner and gestures used by the subject. What specific words were used by the subject?
5. The interviewer must ask the right questions so they receive information that will be helpful in making accurate threat assessments and identifying potential risk control actions. The reviewer should ask about other possible related incidents such as vandalism or other property damage (e.g., slashed tires), following behavior, unwanted/unsolicited flowers or gifts, harassment by telephone, e-mail, voice mail, letters, notes, tapes (audio/video). Has the individual engaged in weapon talk or other action related to weapons?
6. Assess the impact of the behavior or communication on the collateral and assess the level of emotional distress including the operationalization of distress. Establish any known secondary gain for distorting the report of an act or communication by either party. Has any event occurred that could cause an increase/decrease of frustration or emotion such as an upcoming performance review?

Timely collateral interviews are essential, especially when it becomes apparent from the recipient interview and supporting documents (i.e., letters, faxes, other types of communication) that the individual may be escalating and may pose a higher risk.

By conducting detailed collateral interviews, the interviewer will be able to formulate a clear picture of the two parties and their interaction. Collateral interviews will give the investigator an idea of how the alleged perpetrator has responded to boundaries and limitations in the past. This assists in the evaluation of risk control options that may be tried in an attempt to curb inappropriate behavior. For example, if collaterals say that the individual has previously been served with a no trespass order or a restraining order, yet continued to violate those orders, it is likely that the same behavior will occur if this boundary is used again. To reduce the likelihood of this boundary violation occurring again, some new risk control action needs to be considered. Therefore, the investigator will have an opportunity early in the investigation to formulate a swift response to that boundary probe.

ASSESSMENT OF CASE MATERIAL

Based on the need to first assess and then manage a potential threat, the interview of any individual in question is a process. It is important to first organize the available information and to make a list of grid points based on the information provided by the recipient and/or collateral sources. This grid should provide a clear set of discrete behaviors, communications, and incidents, which represent the information that will go to the threat management team and be used by the reviewer to conduct the interview with the individual in question. This grid represents a distillation of all of the information into the relevant data for consideration by the organization in determining whether a policy violation and/or a risk exists that will drive a review by the organization. The data from these interviewers as reported to the team are not the sole sources of information for personnel decisions. Personnel decisions will require other independently gathered information. For example, human resources may be asked to review the company response to similar situations in the past. This type of information is not directly relevant to the threat assessment.

The information grid assists the interviewer in giving the individual, by the end of the interview, the key issues of concern. The individual has the opportunity to respond to each of those issues before they are presented as a part of the assessment to the threat management team. Additional individuals may be identified that require interviewing. The case material will also provide some clues to the manner in which the reviewer needs to approach the interview. This will include collecting additional information from case facts as well as the individual's reactions and responses to already existing information. Additional collateral sources may be identified who can corroborate explanations of alleged behavior.

Before interviewing the individual in question, construction of the grid needs to occur and any missing pieces of information evaluated to determine whether the interview can proceed without the missing elements. Often interviews can proceed and missing information can be obtained later. The use of the grid will quickly make apparent the importance of each piece. In a case of a communicated threat, it is important to review the information associated with the threat and to engage in a thoughtful review of the available data. Fein and Vossekuil (1998) have offered a set of questions that help to thoughtfully review a communicated threat in a case of possible targeted violence. The questions provided later in the chapter are helpful in organizing the information collected in the review. A review of these questions with case relevant data provides insight into the level of risk in the situation based on identified communications, behavior and activities (Fein and Vossekuil, 1998).

Individual in Question Interview Guidelines

The following guidelines can be used when the decision is made to conduct the work behavior-related review with the individual whose behavior, actions, or statements have come into question. They do not restrict or replace the judgment of the threat assessment team or interview professional at the site.

When conducting the individual in question interview, the reviewer has several objectives. The investigator will want to demonstrate to the individual that his or her behaviors have been noted by the threat assessment team. The reviewer will be evaluating the subject's reactions, gathering additional information to show corroboration, and identifying other possible persons with risk. The primary goal of the interview is to give the individual the opportunity to tell his or her story, and respond to the grid points before they are presented to the threat management team. Subjects often have a strong need to tell their stories in a receptive and respectful setting, and quite often the reviewer is the first person to demonstrate any interest and the first person to listen in detail. Therefore, the first objective is to facilitate a dialogue with the individual and to listen to his or her version of events related to the allegations. To do this in a respectful manner encourages the subject to share thoughts, feelings, and, possibly, plans.

The second objective of the interview is for the reviewer to disrupt forward motion as it relates to the potential for escalated risk and encourage a change in the individual's behavior. As well, the reviewer is aware of points of behavior by the organization that are increasing the risk of the situation and can be changed. The reviewer will be communicating that certain behavior is unwanted, unacceptable, and needs to cease. Finally, the reviewer must communicate that if the communications and behaviors continue, consequences will result within the organization and possibly outside.

It is generally recommended that if an interview has been determined to be necessary or useful, it should not be conducted too early in the process. Generally, collecting collateral information is the first step when deciding whether to interview. The investigator must try to assess how an interview may impact the individual. For example, could an interview of the individual increase his or her interest in the recipient or increase risk potential action? The subject may be feeling a certain degree of desperation that contributes to the motivation to take action. Therefore, it is critical to assess how desperate he or she may feel, and how close he or she is to choosing violence as a solution (Fein, Vossekuil, and Holden, 1995).

Based on events that transpire in any case, such as an employment action, a court hearing, an interview by law enforcement, or a change in some other aspect of the subject's life, the subject may have a perception that time is "running out." Careful attention should be paid to the possible behavioral response prompted by the interview, e.g., "before they put me away." Organizations are often fearful of talking to the individual about the behavior, fearing it will make things worse. This is in most cases the very action that needs to be taken. No one has spoken directly and clearly to the individual about the behavior, actions, and/or statements and set limits. Working indirectly heightens the individual's concern that something is going on. Facing these situations directly allows for a thorough assessment of the threat and risk.

The following points should be made at the beginning of an interview:

- The interview is not confidential but on the record.
- The interview is not intended as a medical, psychological, or psychiatric evaluation.
- The interview is not an appropriate time to share private medical or mental health information.
- The interview is focused on work-related behavior, actions, and statements.
- The interviewee may take a break at any time (rest room, phone, smoke).

Employee Interview Sequence I

1. Review preinterview physical search issues
2. Management/HR introduces employee to interview

- Management conveys material orally or in writing
- Consultants are used when company has problem in production or service that is difficult to solve (e.g., problem(s) with employees working together, a human resources issue, and/or a team conflict)
5. Consultant company is:
   - Consultants are (first and last names)
   - As necessary, consultants have backgrounds in organizations and/or psychology and/or risk assessment

Employee Interview Sequence II

1. Introduction to employee by interviewers
   - Introduce by name
   - Ask what employee has been told about purpose of meeting and why consultants are present
   - Provide same explanation as provided by management/HR
   - Explain that all information that consultants receive from employees is provided to management
   - Document that this information has been conveyed
   - Provide individual with "Your Interview Today" sheet to sign
2. Transition to interview
   - When did you first join the company?

Employee Interview Sequence III

1. Conduct brief work history of employee prior to company
   - How did you come to work for this company?
   - What company were you working for previously?
   - What dates (approximate) did you work for that company?
   - What type of job did you have with that company?
   - Complete this inquiry for at least the past three jobs or past ten years, if possible
2. Conduct complete work history of employee at company
   - How long did you have your first position at the company?
   - What unit or work site were you assigned to?
   - You reported to whom?
   - When did that job position change?
   - Why did that change occur?
   - What was your new position?

- Who did you report to in this new position?
- Duplicate the above list of questions until you have completed the work history of the employee up to the present.

3. Transition to current work unit
   - What are the positive things about working at this company?
   - No matter how positive the workplace might be at a company, there are almost always negative parts. What are the negative things about working at this company?
   - What needs to be improved? (Allow employee to determine meaning of question.)
   - What difficulties, if any, have you noticed among co-workers?
   - How would you view yourself as an employee (positive and negative aspects)? Each of us as an employee has both positive characteristics and areas for improvement. What are the positive characteristics about you as an employee? What areas for improvement do you see for yourself as an employee?
   - Are there factors that impact your job performance?
   - We have talked about some of the difficult parts of doing the job at this company. From time to time, almost all employees have things that happen outside the job that can affect their ability to do the job. How do you think the company responds when an employee has an outside problem that affects his or her ability to do the job?
   - Review with labor counsel these questions depending on the nature of the case: Have you had any stresses that have affected your ability to do the job at the company? (If employee begins to discuss health/medical issues, stop and remind the employee that this is not an appropriate time to share detailed information.)
   - How did those stresses or problems affect your work?
   - How could the company have responded better when you had those stresses or problems?

Employee Interview Sequence IV

1. Primary effort is to give employee the opportunity to initiate reports of incidents
2. If employee does not report incidents, use grid to review issues
   - "I have heard something about . . ."
   - "I am unclear about . . ."

3. Each incident on the grid needs to be discussed with the employee
   - Employee's factual account of events
   - Employee's view of co-workers' reactions to the behavior/threats
   - Employee's view of management's reaction to the behavior/threats

Employee Interview Sequence V

1. Determine if employee acknowledges behavior/threats
   - If employee acknowledges behavior/threats, seek explanation of how this occurred
   - If employee acknowledges behavior/threats, seek explanation of how this could be prevented from reoccurring
   - If employee does not acknowledge events, ask how the information could have been misinterpreted or falsely reported
2. Have employee respond to reported differences or inconsistencies in grid points
3. Inquire about gun and weapon ownership, also about weapon–talk in the workplace
4. Inquire about involvement with the police within the last three years for assault/threat behaviors
5. Inquire about prior behavior/threat situations

Employee Interview Sequence VI

1. Engage employee in problem solving
   - How can this situation be resolved?
   - Role reversal: What would you do if you were supervising someone and this situation happened?
   - Review range of options that management might consider in resolving this situation; note carefully both the verbal and nonverbal response
   - What do you think management is likely to do?
2. One interviewer exits to contact coach, if appropriate
   - Brief coach on category designation and data that justifies category designation
   - Coach provides specific areas and/or questions for further exploration
   - Interviewer returns from telephone call with coach; asks additional questions

3. Interviewer quickly communicates category classification using a number in the interview with the interviewee present
4. One interviewer exits to conduct briefing on category classification
   - Security team first
   - Threat management team second, if available, or team representative
   - Exiting team member may or may not return to the interview
5. Management enters interview room if the situation is safe, asks remaining interviewer if interview is complete: remaining interviewer exits
6. Management concludes predetermined management action (continued administrative leave, return to work, termination, disciplinary action), informs employee of any restrictions on entering company property, calling or writing to other company employees, and notifies employee of point of contact within the organization as well as time frame of organization response
7. Management/security follow predetermined plan to exit the individual from the facility, if required. Make sure individual actually leaves the parking lot. Ensure that employee does not return to the parking area or building
8. Interviewers in private make final determination of category classification and supporting points
   - Assign areas of responsibility for report to management
   - Interviewers transcribe the category classification and supporting points
   - Initiate and complete telephone conference call with team coach, if required
9. Conduct management briefing
10. Follow up as necessary

THE INTERVIEW ENVIRONMENT

Setting

When considering an interview it is important to give some thought to the setting in which the interview will take place. The setting should be nonthreatening, ensure some privacy, and be respectful to the individual. The setting should also provide adequate safety for the interviewers. The interviewers' attitudes need to be respectful and communicate an interest in listening to the individual's side of the story. Obviously, those who refuse to answer questions or request an attorney are very quickly conveying that they perceive something is wrong and may feel a sense of persecution. This should be noted and documented. The threat assessment team needs to have reviewed the issues related to representation in the interview and organizational policy. In a unionized environment, the employer will have a representative present in most cases as the information may lead to employment consequences. Even if the individual refuses union representation, it is advisable to have a representative present to hear the refusal and to be accessible in case the employee changes his or her mind.

In the case of a nonunion employee, organizations need to review with counsel their interpretation of the National Labor Relations Board ruling on representation and organizational policy. In general, it is not advisable to have attorneys in such interviews, especially if the interviewers are not attorneys. Each situation needs to be reviewed and discussed, and responses formulated if the issue arises, which it will not in many cases.

This is a good point to discuss an issue that runs through the interview process and beyond. The organization must project to all individuals involved that this is a well-considered, organized, and reasonable response to the situation. This projection assists in bringing order and boundaries, which helps contain behavior and limit risk. Chaos and disorder create an environment in which aggressive individuals find not only justification for their behavior but opportunity.

Plan

Develop a plan before the interview that includes consideration of the following: security, safety, setting, dress, approach, introductions, central questions, and information grid. Plan an opening statement, such as "I am here with my associate to gain a better understanding of the circumstances." "It appears that something important needs to be communicated; others need to understand." "We are here to listen and understand."

In all cases, reviewers conducting field interviews need to do so in pairs. Becoming aware and remaining sensitized to safety issues is paramount, and interviewers must not let their vigilance down because they feel some degree of rapport. Overreliance on rapport in the face of communications and behavior that suggest the subject may pose a threat can, and has, resulted in tragic consequences for many law enforcement professionals throughout the United States.

Prepare

The interview needs to be well prepared but not scripted. Over-scripting an interview can often lead the subject to believe that the interviewer lacks genuineness, flexibility, and relatedness. In an interview of this type, rapport is necessary to create an opportunity for the subject to feel comfortable sharing his or her ideas. This may be the first time that anyone has provided an opportunity or the time to listen to the subject's story. The investigator will have an organized information grid and timeline to use as a format if structure is needed.

INTERVIEW PHASES

Phase I

During the course of the interview, the investigator is collecting information and conducting an assessment, documenting in detail the initial and subsequent contact/incidents/communications, clarifying in detail the motivation or intent of each communication and/or behavior. In the initial phase of the interview the investigator remains subject centered.

The focus of the interviewer is on understanding some aspects of the individual and his or her work history. Also, this is the opportunity to explore the relationship with the organization and the work environment. The roles of supervisors and the general interface with the organization is explored. The objective is to develop an understanding of the broader context of work within which behaviors, actions, and statements may have occurred.

Some individuals will want to rush to the incidents that they believe are the problem. It is important to set a pace and control the interview. Slowing the process allows the individual to organize the responses as the interviewers provide a structure to the process. The structure also assists in communicating that this is a process that the interviewers are familiar with and comfortable performing. Finally, the structure tells the individual that the workplace experience is important in helping the interviewers understand what has happened and in bringing meaning to the events. This context can then be used to convey the individual's perspective to others.

Phase II

In phase II the interview is focused around the issues raised by the subject. The interviewer listens to the rationale for engaging in the actions and communications that have been alleged. This is not the point in the interview where the investigator reviews potential consequences; this is the information gathering and assessing phase. During this phase, the investigator should try to maintain the focus and the pace of the interview, especially if the subject begins to stray away from a particular issue. Using the need to document issues on behalf of the subject, the investigator can redirect the focus of the interview: "Wait a moment. I need to get this down." "What you're saying is important." "I want to be able to understand what you are saying." "I recognize that that is important, but I want to completely understand this point first."

After capturing a comprehensive view of how the subject explains the allegations, it is important to assess to what extent he or she appears to be fixated and focused on the target and how close he or she may be to taking action. It is also critical to try assessing a time imperative, if one exists, to be able to react to planned behavior that could result in a tragic outcome.

At this point, the interview shifts from a more collaborative effort to one where the reviewer takes greater control. The interviewer begins to present to the subject what has been said about the subject's communications and behavior. The presentation of new or contradictory behavior is not done in a confrontational manner, but tends to keep to a theme of clarification and problem solving. This suggests that the reviewers have data that may be inconsistent with what the subject has offered.

Phase III

Out of respect for the subject, and in the spirit of the interview process, resolving the inconsistencies or obtaining collaborative information to support a contrary view is important. This is also a time where the subject is provided an opportunity to view his or her actions as misperceived by the target. This is especially important when there are contradictions and inconsistencies. "How could others have misinterpreted your intentions?" "Perhaps they don't fully understand based on what has been communicated." "Are there additional details that we have not discussed that might affect how others perceive the situation?" This also provides an opportunity for the subject to offer further disclosure: additional information, more details, or the involvement of other individuals not previously mentioned. For example, we participated in an interview with an individual who was under review for stalking and making threatening statements regarding employers. During the interview (as part of a condition of his parole) he disclosed that while he did make statements of harm to his former employer, he was not going to kill him. In reviewing his previous employment record, however, he mentioned another supervisor whom he stalked and for whom he made more detailed plans for violence. There are numerous instances when subjects have more than one target, and the investigator must thoroughly explore any further disclosures by the subject. Determining whether others may be at risk in the organization is a concern.

After the reviewers have explored the issues surrounding the communication or behavior and have a reasonable degree of confidence in their assessment of the risk situation, the reviewers should shift to discussing with the individual the potential consequences of the behavior.

The range of actions that an organization might take needs to be reviewed with the individual. In most cases, the management team has not made an employment decision yet, or that decision is being made while the interview is ongoing. Therefore, the full range of potential options needs to be explored from doing nothing to termination and the levels in between. This may be the first time the employee faces the possibility that serious employment consequences may arise from the situations. In other cases the employee assumes termination and has not considered that the organization may have intermediate steps that could occur if behavior is changed.

Chapter 7

Liaison with Law Enforcement

Over the past ten years significant modifications in laws addressing communicated threats have changed the landscape as it pertains to responding to threat in the workplace. Based on these legislative changes, law enforcement has taken a more active role in investigating and responding to threats in the workplace. In many police departments, officers and investigators have dedicated new efforts to resolving communicated threats in the community. There has been a significant increase in police training as it relates to law enforcement and in publications dedicated to providing guidance to law enforcement in investigating and managing communicated threats. This includes specialized training by the U.S. Secret Service, National Threat Assessment Center, The FBI National Academy, and the National Sheriff's Association. A national organization was originated by the Los Angeles Police Department's Threat Management Unit: the Association of Threat Assessment Professionals (ATAP). ATAP holds an annual conference that provides basic training and advanced seminars for police. As a result, law enforcement has developed strategies to assess the increased number of threat assessment cases in the workplace.

It has become easier in some jurisdictions for legal, human resource, and security professionals to obtain investigative support from law enforcement. Today is very different from years past, when corporate professionals did not expect a response from the police or from the judicial system until some overt act occurred. The police are more organized and more responsive to organizations regarding workplace violence, and organizational professionals should develop contacts and foster relationships with local law enforcement. The open line of communication expedites communication between organizations and law enforcement during times of crisis. The corporate professionals should develop an understanding of what type of information is most useful to law enforcement and to provide such information efficiently. Corporate professionals are more likely to receive assistance when they present their case to law enforcement in a manner that is most consonant with their methodology and philosophy. This includes reporting facts and statements that reflect elements of potentially criminal acts. Such presentation will facilitate law enforcement participation and the greatest cooperation from the local district attorney. The local district attorney presents a case in the interest of the people to a judge who will outline certain limits and boundaries that, if violated, will result in criminal penalties.

In addition to those cases that meet the criteria for criminal code, the organizational professional should develop a relationship with law enforcement for the purpose of support in regard to security and a community policing response to a potential threat to the workplace. In many cases, where employees have been terminated or who pose threats to a company from the outside, the communication with the police is critical. It is not to collaborate on investigative efforts, but to get attention and response from law enforcement on cases that may escalate. Often these cases start out as not prosecutable but a law enforcement presence in the form of police on patrol would be useful as a form of risk control. Prompt police reaction when called in response to a boundary probe by a former employee can be useful in reducing risk. Police should be cognizant of individuals who are targeting employees as a result of a domestic violence case or a stalking situation; these individuals may follow the target to the workplace.

The overall goal of a police liaison is to develop an understanding of how the law enforcement department in your jurisdiction approaches such problems. A second goal is to develop interactive contacts between your organization and someone higher in the police ranks than patrol officers who respond to calls. A third goal is to familiarize the police to your organization, location, and planning for such threat incidents. A fourth goal is to develop a long-term relationship prior to the need to access law enforcement resources. By becoming a partner in your community with law enforcement and treating them as a valued source of expertise rather than just muscle, you enhance the likelihood of communication, understanding, and action when you need support. Law enforcement agencies have many competing demands on their resources; actions that facilitate the partner also enhance the likelihood of your ability to access resources.

Employment reviews and workplace investigations conducted by private security and the human resource professionals will inevitably differ from a more formal police investigation. However, similarities exist. They both gather information from victims, collaterals, and, in many cases, the subject. Therefore, the accurate recording of specific communications and observable behaviors are of utmost value to the police, who may have a concurrent investigation on the subject. For example, an employee who commits domestic violence or violates a restraining order may have broken the policies of the company as well as the state or federal statues. On the other hand, policy violations may occur with no comparable legal statute violation.

A formal law enforcement investigation should not limit the organization's review. The organization has different objectives, outcomes, and considerations from law enforcement. A well-established liaison will allow the organization, if necessary, to coordinate with law enforcement the timing of separate investigations. Corporate counsel can help guide conversations with law enforcement agencies as well as ensure organizational and employee cooperation guidelines are followed.

The assessments and analyses that are conducted by organizations may be incorporated into a management plan, which may be supported by limited law enforcement. This model differs from assessment models where private sector threat assessments conducted by security or human resource personnel were supported only by management. Assessments by organizations with the collaboration of enforcement support can facilitate action by the judicial system. Furthermore, law enforcement officials have access to more detailed information than private sector personnel, such as arrest records, motor vehicle records, and driver's license photographs, which augments the corporate management team's ability to manage a potential threat. Conversely, employers' access to information is limited by employment laws and regulations (e.g., restricted access to medical information by the Americans with Disabilities Act), yet employment records often contain valuable information regarding work-related behavior that is specific to the context in which the communicated threat has occurred and where the patterned behavior assessed for risk exists.

The primary methodological difference between threat cases and other crimes is the paucity or absence of physical evidence in threat cases. Although some threat cases contain physical evidence (e.g., letters, e-mail, pictures, or audiotape), many cases only offer witness observations of the communicated threat. The challenge posed to reviewers is to accurately and comprehensively collect subjective evidence of the alleged perpetrator's behavior and communications. Therefore, interviews with the recipient of the threat, witnesses, collateral sources, and the perpetrator are critical in the collection of data. The partnership in more serious cases between the law enforcement and the organization regarding a threat assessment becomes important in facilitating the judicial process that has sharper teeth in setting limits and boundaries for an employee recently terminated or for an outside individual who has communicated threats against a company.

Case Example

A forty-six-year-old man made troubling comments during a training session regarding his alcoholism. His comments reportedly caused such a disturbance that he was counseled by his supervisor and referred to the employee assistance counseling program. In addition, on two separate occasions, he displayed inappropriate behavior by storming around the office, cursing, and throwing objects. During an employee workshop, he made several inappropriate comments in an attempt to disrupt the class. In both instances, his supervisors counseled him and asked him to leave the office for the day. After two months of stress leave, on his first day back in the office, he engaged in a verbal outburst during a meeting. He subsequently requested a transfer.

His supervisors documented a pattern of unusual agitation on minor issues, including unreasonable complaints about his job, inability to perform acceptable work, and allegations that his co-workers were conspiring against him. Specifically, they noted that the EAP counselor suggested he receive treatment for an alcohol addiction. However, he denied that he had a substance abuse problem. Furthermore, supervisors noted that he accused them and human resource personnel of conspiring against him.

The individual was subsequently twice hospitalized on a voluntary basis for homicidal ideations. He was treated for psychosis, as well as suicidal and paranoid delusions associated with his co-workers. Essentially, this individual believed that his supervisors and co-workers were participating in an organized conspiracy to discredit him and drive him from his job. His physician recommended disability retirement.

Over the course of three days, the subject began to leave harassing voice mail messages for a co-worker. One of his messages stated, "Hi Diana. It's Bill. Just wanted to say Happy Thanksgiving. And you give this message to Yvonne. Tell her if she had been off the property the day she hollered at me, I would have beat her mother-f****g ass. Bye, Diana." During the following month, the subject's disability pension was approved. He was diagnosed with delusional disorder, paranoid type.

It was not noted until much later that his retirement papers contained disturbing comments. For example, one of his statements regarded a meeting he had with a human resources specialist: "I started to grab her by the throat and choke her until the top part of her head popped off. Then I was going to step on her throat and pluck her bozo hairdo bald, strand by strand . . ."

Approximately five months later, the subject began to call a former co-worker, telling this individual that he was following a former supervisor and her family. He provided specific details as proof of his actions. In addition, he commented that he knew where some of the targets lived and the types and colors of vehicles they drove. The individual also made personal comments about the target's family members. He further stated that he had three guns for each of his former supervisors.

At this point, law enforcement officials were notified. The law enforcement investigator concurred with corporate counsel who recommended to the targets that, among other security measures, they should obtain restraining orders. The five targets declined to do so, stating that they believed restraining orders would only agitate the subject.

During a subsequent counseling session with his therapist, the subject threatened to harm the five female former co-workers, including the human resources specialist. The therapist provided a warning to law enforcement and the named employees. A threat assessment was conducted, which involved analyzing the subject's letters, voice mails, and interviews of various individuals. The threat assessment revealed several violence risk factors, including past violent behavior, heavy alcohol abuse, and not taking prescribed medication. In addition, the subject's communications displayed organization and contained specific threats to harm others. He wrote, "Don't let the passage of time fool you; all is not forgotten or forgiven," and, "I will in my own time strike again, and it will be unmerciful." Based on the available information, the situation was assessed as containing significant risk, category 2. The available data suggested that the subject was escalating. Specifically, he was becoming increasingly fixated and focused on the targets. His communications articulated an action imperative that suggested an increasing risk for violence. It was determined in cooperation with local law enforcement that investigators would conduct an interview with the individual. In order to conduct the interview, the interviewer developed strategies to approach the nonemployee.

A report authored by the threat assessment group was provided to the field investigators and the district attorney. It was also used in subsequent judicial proceedings. The report outlined the subject's patterned behavior, his progressive escalation, and his continued fixation on the targets. It addressed specific aspects of the subject's behavior, including his disruptive behavior in the workplace, continual threats to co-workers, and the following and harassing of his former co-workers. The report also provided recommendations such as strategies for approaching and interviewing the subject. Comprehensive security recommendations for the targets, including general safety guidelines and procurement of restraining orders, were also discussed. The report was written in a manner that law enforcement could easily follow, interpret, and incorporate into an investigation worthy of prosecutorial merit.

The goal of the law enforcement field interview was to develop sufficient rapport with the subject. By telling his side of the story, the subject provided valuable assessment data on his closeness to an attack, as well as factors that could increase or decrease a potential attack. During the interview, he assured the field investigators of his intentions to pursue legal reparation. After obtaining additional assessment information, the investigators advised the subject of specific boundaries (i.e., no contact orders) and consequences for violating them (i.e., arrest and incarceration), should he continue his threatening behavior and communications. In response, the subject convincingly reiterated his intention to pursue a legal resolution.

Four months later, the subject mailed numerous letters to the five targets, stating that he wanted to "execute" one of the women. Field agents submitted the letters for an updated assessment. Based on the ongoing assessment and insight into the subject's thought and behavior processes, the data indicated that the subject was potentially close to a violent action. In response, a conference call was arranged with the district attorney. During the conference call, the investigator and threat assessment professional articulated aspects of the threat assessment. The threat assessment professional provided an assessment of the situation risk for violence, while the investigator presented evidence of law violations and law enforcement actions taken to date. Through the threat assessment report and supporting evidence, the district attorney was able to obtain arrest and search warrants. The threat management team supported the subject being expeditiously arrested and held without bond. Based on their knowledge of his past patterned behavior, the investigators/police designed a safe manner of approaching the suspect. He was successfully arrested and held without bond. Six months following his arrest, the subject was found not guilty by reason of insanity.

The primary concern for any corporation when it comes to a threat situation crisis is the safety and security of its employees. During a threat management crisis, it is often helpful to provide a comprehensive security briefing to any victim or witness to a workplace violence incident. Police community liaison officers are one of the best resources for such briefing. Police have experience in working with citizens on such issues and a knowledge of special community risk issues.

A police contact card is a tool that facilitates a law enforcement response. An employee who is subject to violence/threat of violence/

harassment/emotional distress may be subject to contact by the individual of concern. This contact may take place outside of employer buildings, on employer grounds, or off work. Such contact may result in a call for law enforcement response. The employee should be able to communicate effectively with the responding patrol officers. Responding officers are likely to be faced with an employee who is attempting to orally communicate a complicated history of past events in which the employee is the victim and the other party is the offender. Similarly responding officers are likely to get a similar story from the offender with roles of victim and perpetrator reversed. The goal then is to assist the responding officer in quickly and effectively determining that the employee is the likely victim and the other party is the likely offender.

Accordingly, we recommend preparing a plastic laminated card, approximately 3 × 2 inches, that provides the responding officers at a glance with all of the preprinted information likely to be needed for the police report they will subsequently prepare. The plastic laminated cards should be placed in relevant but secure locations: locked car storage compartment, wallet, purse, etc., where they can be readily produced if necessary.

Side one of the card contains data about the employee. Side two contains information about the potential offender. This may vary depending upon the status of the potential offender: employee on administrative leave, terminated employee, third party/nonemployee. This general recommendation should be reviewed and approved in accordance with individual state law by employer general counsel and/or a labor law attorney prior to implementation (see Appendix A).

Chapter 8

Analysis of Threats

Communicated threats can be expressed in a variety of modalities by anonymous or known authors. A communicated threat may be written in the form of a letter or electronic mail message. It may be recorded in a voice mail, audio, or videotaped message. In some instances, threats are orally communicated and witnessed by a co-worker or witness(es). In some cases, the communicated threat is not a verbal message but rather a demonstrated behavior, such as a gesture or action that leads to emotional distress for the recipient. In this chapter we shall deal primarily with communication that does not occur in person. These communications may encompass a single message or multiple messages over time and use various delivery methods. They may range from serial messages left on portable toilet walls at a construction site to increasingly intense letters sent to the CEO of a corporation.

In any form a communicated threat requires a structural framework to assess both the level of risk and the risk control actions that need to be considered. This structural framework serves to guide the flow of information into a useful form and helps decision makers determine whether the level of threat has changed and whether a review of risk control actions is needed. In any communicated threat it is critical to try and determine the intention of the individual who communicated the threat. It is also critical to assess whether the individual is making a threat, poses a threat, or is somewhere in the process of moving from one to the other (Fein and Vossekuil, 1998).

The communicator may be known, and collateral witnesses may provide additional information regarding intent and relevant risk factors in such circumstances. If the communication is anonymous, additional information regarding context and circumstances is generally not available.

A communicated threat for any organization is quickly characterized as a concern. It may be the first signal of a potentially escalating process that might result in a violent action. It may be a communication that reflects an individual's frustration, loss of judgment, or impulsivity. It may be a communication that is consistent with an individual's attempt to influence an action of the organization. For the organization, it is important that the threat management team be able to rapidly determine whether a communicated threat requires immediate action or becomes a part of a more timed response. The communication may represent an expressive discharge of anger that is viewed as a solution to the individual's pent-up frustration and dissatisfaction with the work, business, or customer relation's issues. In other cases, individuals who communicate threats may be providing advance warning of action they plan to take. The goal in such instances is to increase the fear and suffering of individuals in the organization in anticipation of the action.

In this scenario, the individual has communicated ideas regarding threats of action or actual action as a possible solution. Over time the individual may begin to feel that communicating a threat has failed to create the desired action or impact a work-related situation. The individual begins to experience responses that merely communicating a threat was unsuccessful and that action is imminent and a violent act is a solution. Some communications may or may not pose a threat but are perceived by the receiver(s) as distressing. The recipient views violence as the proposed solution and communicates distress to the organization, expecting that the organization will do something.

Unlike other employment-related situations where the context in which the communication or behavior under review is known, the communicated threat or observable behavior may not have a context in which the communication or behavior can be anchored and interpreted. Threat assessments and potential risk control actions are often based exclusively upon a review of the materials. Hypotheses are generated that can be helpful in choosing the level and type of risk control actions. Other information not available and/or not reviewed at the time of the analysis may change or alter assessment conclusions and recommendations.

This chapter will outline an approach for evaluating communicated threats for violence potential. This system is considered a descriptive method to assess threatening communications and has demonstrated over time to be useful for providing input to decision maker(s) in deciding on additional actions or implementing a security plan.

PRINCIPLES OF THREAT COMMUNICATION ANALYSIS

Organized versus Disorganized

This principle refers to the level of organization and coherence present in the behavior and/or communication. Does the behavior and/or communication have a central theme? Does the behavior and/or communication keep to that central theme in a logical manner? In cases where the level of organization of behavior and communication is rated as high, the author has stuck to a single theme that is continuous, linear, and logical. In cases where the level of communication is rated as medium, the author has offered multiple themes in the communication. Some authors use related but different themes referenced within one short paragraph. For example, some themes are:

- Racism ("You are the biggest bigot I ever met")
- Retaliation by individuals other than the writer ("If I don't get you first, one of us will")
- Inadequacy of victim's family to protect her ("Your short runt of a father can't help you either")
- Victim taking early cut as a successful resolution of the writer's anger ("You need to take the early out . . .")
- Accidental injury if the victim does not leave with the early out option (". . . before you have an unfortunate accident around these parts")
- Organizational policy decisions ("You apply policies unfairly")

Writers who have become more fully intent on harm to their victims characteristically have a higher degree of specification and organization as to how they are going to harm the victim, and the specific behavior for which the victim needs to be punished. The concern for a victim's physical safety would be heightened if future communications are more specific in their level of organization and action/time imperatives. Increased frequency of communications that specify details of acts/attitudes/personality characteristics of the victim are then more directly linked to acts of harm to the victim (e.g., "I and everyone else knows what you did yesterday and what you have been getting away with for the past three months since you came here, you will not get away with this I promise you, if you don't leave now, you will not be able to walk to work"). Disorganized communication with a multitude of messages jumbled together are often quite distressing to the recipient. From a risk point of view, however, such ideas usually represent a decreased risk for a planned, effective attack. Violent behavior will tend to be quite impulsive. Such individuals draw extensive attention to themselves because of their personal mannerism and disorganized behavior. A complete breakdown in organization often leads to very bizarre and disturbing communication, but extremely limited ability to organize and carry out a planned behavior.

Fixation

The next concept in the analysis of threatening communication pertains to the idea that the author demonstrates a degree of fixation on a part of the organization or the organization as a whole. The individual has concluded that his or her employment and personal problems are the fault of the company. The individual has placed responsibility for the problems on the company. The communication is reviewed for content that reflects blame. This may be due to a recent reorganization, change in benefits, personnel actions, or actions by the organization that are a part of the business process. In situations where the author may not be an employee, the author may be a stockholder who has recently suffered some losses, a former employee, an individual in the community who takes issue with some company action as it relates to the community or the environment, or an individual whose preoccupation with the company is the result of a mental illness.

Focus

The third factor is the degree to which an author demonstrates focus. Focus refers to the situation in which a person has concluded that his or her employment and personal problems are not only the fault of the company, but has concluded that usually one to three specific employees of the company are responsible. The degree of focus reflects the extent to which the author communicates feelings of being singled out by these individuals and these individuals and that these individuals are the source of the problems. The author may also express a feeling of being persecuted. Individuals who are identified as specific targets in the communicated threat are responsible for the employee's difficulties in the workplace. When the author is not an employee, reviewers should determine, if possible, how the targets have become identified and what other situations or interactions may have occurred between the author and the potential targets.

Action Imperative

This principle refers to the need on the part of the person to take personal action to resolve the situation with the company and its employees. Here, the person has determined that all other avenues (administrative, legal, criminal, etc.) are not going to help with the resolution. All other avenues of redress have been exhausted. There is a substantial pressure for the person to take action. Action imperative is evident when the author has articulated a specific plan to carry out the communicated threat. This principle assists in assessing the degree to which the author may feel greater pressure to think about and articulate ideas about a violent action as a solution to the problem.

In cases where the level of action imperative is rated as high, the author has offered considerable specificity in his communication and demonstrates the pressure of words and ideas, which indicate an increasing need to take action. The specific make and model of a particular weapon or method that the author intends to use as an instrument of violence is an example. In addition, a plan that demonstrates considerable detail and reflects knowledge about the daily activities and movements of the victim raises concern to a higher level. Finally, communications that demonstrate through words and ideas that the author has engaged in some form of rehearsal heighten risk. The writer may threaten accidents, which the victim's spouse will not be able to prevent. The writer does not demonstrate the pressure of words and ideas, which indicates an increasing need to take action now. General accident is less specific and more ambiguous. Concern for victim physical safety would be advanced if the following example of action imperative began to appear in future communications or contacts: statements of pressured need for the harm to the victim to occur (e.g., "The company lets you get away with it, you think you can act like this and get away with it, you cannot get away with this, you have been judged, you will be punished just as you have punished others").

Time Imperative

This principle refers to the person's need to take personal action to harm the company now or in the immediate future. Here, the person has not only determined that he or she needs to act to harm the company, but that action must be taken soon or within a near time frame. Time imperative is a principle that reflects the degree that action may be imminent. The author demonstrates a certain degree of pressure in words and ideas that reflect a need to take action. Time imperative is related to the articulation of an action imperative. If the subject has articulated a plan it is important to assess if the plan could be realistically acted on within the time frame specified in the time imperative.

Concern for victim physical safety would be advanced if statements of pressured need for harm to the victim to occur very soon began to appear in future communications (e.g., "Your time is at hand, this can go on no longer, soon you will reap what you have sown, you must leave now or you will be responsible for what happens next").

COMMUNICATED THREAT RATING

The threat analysis model, which assesses a communication based on organization, fixation, focus, action imperative, and time imperative, is designed to be used in conjunction with the threat assessment model. The use of a model offers organization to data that guides the ongoing threat assessment. Rating communications across all of the factors help develop a baseline for initial decision making of appropriate risk control actions. This baseline then becomes a marker if future communications signal that something has changed in the situation. Each rating reflects both the degree of emotional distress and physical harm that may be experienced by the recipient. Each factor is rated independently. One rating does not necessarily outweigh any other. However, if a communication is evaluated according to the degree of pressure an author is experiencing across a continuum of idea to action, the obvious presence of an action and time imperative combined with the other factors (organized, fixed, focused) leads to the consideration of a more aggressive security and law enforcement response.

An author who sends a poorly organized communication that reflects some fixation on a company, no focus on specific individual, and fails to identify a specific action plan or time imperative suggests that the potential for action at this time is low. Although the communication may be rated as having some impact on the recipient's experience of emotional distress, the potential for physical harm appears low ("If the company fails to stop its job and profit loss, something is going to happen that everyone will regret"). In comparison, a more tightly organized communication that contains very specific themes relating to fixation, identifiable targets associated with focus, and a well-articulated action plan and time frame suggests considerably more pressure in words and ideas in regard to the potential for action. The degree to which the recipient experiences emotional distress would be rated as high as would the potential for physical harm.

> Company XYZ has continued its manufacturing process that places the environment at risk. Since Mr. Jones is clearly in charge and refuses to make changes, he will have to suffer. I have tried to get others to recognize the terrible impact to no avail. Now that the time of your stockholders meeting is approaching, Mr. Jones must be removed by force. Specifically, he will be shot to death—an execution, not a murder.

It is important to rate a communication in regard to both emotional distress and physical harm, as risk control considerations for both need to be addressed. These dimensions may have employment as well as legal implications. The degree to which physical harm becomes a concern may drive additional law enforcement and security interventions as well as the speed with which they need to be implemented.

Before any management of an assessed communicated threat is recommended, it is useful to evaluate the threat in accordance with the IAS category system. Behavior by employees and/or former employees at risk for violence/harassment are most usefully classified into one of the following five IAS categories:

> *Category 1:* Available data suggest high violence potential; behavior suggests immediate arrest, hospitalization, or major organizational response. In this category, the individual meets the state penal code criteria for immediate arrest and/or state criteria for involuntary psychiatric hospitalization for danger to self or others.
>
> The unknown writer of these communications is not known at this time to have demonstrated violent behavior in the past, which would have qualified for inclusion in category 1.
>
> *Category 2:* Available data suggest high violence potential; behavior does not suggest immediate arrest, hospitalization, or major organizational response. In this category, the threat of violence is accompanied by a quid pro quo to the threat, which means that the individual is going to hurt someone if some designated event does or does not happen.
>
> The unknown writer of these messages is not known at this time to have demonstrated potentially violent behavior in the past, which would have qualified for inclusion in category 2.
>
> *Category 3:* Available data suggest insufficient information for violence potential; sufficient data support perception of knowing infliction of distress on co-worker/others. In this category, it is the threat of violence rather than the act of violence that is important. The threat of violence is intended to cause co-workers or supervisors sufficient distress so that no employment or interpersonal action would occur that would be adverse or intrusive to the employee making the threat.
>
> The writer of these communications would characteristically intend to cause intentional emotional distress upon the victim co-worker.
>
> *Category 4:* Available information suggests insufficient data for violence potential; sufficient data support perception of unplanned infliction of distress on co-workers/others. In this category, the threat of violence occurs and is of such a nature that it could reasonably cause emotional distress in co-workers. The individual does make a single threat or threatening

behavior, but is unaware at the time of the impact of the behavior and does not have the intent or motive to cause distress in co-workers.

Writers of communications such as this case have given thought to both the construction of the content of the note and to the means to get the note to the victim without becoming identified. Such behavior would then not qualify a writer for category 4.

Category 5: Available data suggest insufficient data for violence potential; insufficient data of behavior resulting in distress in co-workers/others. In this category, insufficient evidence is present for either violence potential or infliction of emotional distress. Available information indicates that the employee currently under evaluation may be the object of an unfounded allegation of violent threat or infliction of emotional distress by another co-worker for as yet undetermined motivations. Since the identity of the writer is unknown, the behavior would not qualify for category 5.

CASE EXAMPLE

- Package #1 was sent to ABC Studios, Somewhere, USA, December 1998 and forwarded to 456 Studios
- Package #2 was also sent to XYZ Studios, Somewhere, USA, received in February 1999 and forwarded to QRS Studios
- Package contained portraits of Jane Doe

Several drawings accompanied written correspondence and enclosed objects. Native American flourishes abound. Each drawing characterizes the target, Doe, as a stylized, fantasized "Indian Princess." She is draped in an Indian-style dress featuring fringe along the bust line. A campfire with colored flames is prominent in the background of the more detailed drawings. One drawing features Doe, naked, seated before a fireplace, wrapped in what appears to be a buffalo skin robe with "lover's wine glasses" prominently displayed on the fireplace hearth. An accompanying letter spends two paragraphs expressing

want[ing] to tear your clothes off of you and stop imagining just how lovely your breasts are and how they would feel cupped in my palm with their warmth filling my hand as I press your (lips?) torn (torin?) next to mine and melt together and take your sweet lips and fill you as I feel your warmth and silk around me ... what I hear is more than just sex ... that's nothing, rutting is rutting ... [the fantasy continues]

Objects Accompanying Correspondence

Three objects were delivered with the second set of correspondence received in February 1999.

Item 1 appears to be a man's engraved silver ring with simulated turquoise chips in a "thunderbird" image. The image itself is generally associated with Hopi and Navajo iconography. However, it does not appear to be an authentic Native American production, but rather of amateurish construction such as that seen in an institutional jewelry-making class (adult school or occupational therapy [OT] class). An OT class would suggest its origin as a hospital or prison setting. Decorative border markings appear on various surfaces but are either exclusively decorative or have highly idiosyncratic meaning to the ring's maker.

Item 2 appears to be a personal hunting or personal power "fetish," also probably not constructed by Native Americans. The braided lock of hair attached to the object may be human hair or animal hair and could be forensically evaluated if certainty is required. These fetishes are sometimes attached to a favored hunting weapon, usually a rifle, and stand as a symbol of both the hunter's spiritual commitment to the hunt, as well as a divine request for spiritual guidance in concluding a successful hunt.

Item 3 also appears to be a "fetish"; however, it holds no known significance. The gray cockle or clamshell appears to have had some type of marking or writing on it but is indecipherable at this time.

No direct reference to these objects was found in the accompanying written correspondence. However, a passage that clearly appears to be written in a Native American language was penned in the margin of a document and says, "You two face backstabbing haughty 'aksinee' 'puh no kahm 'thh' would not recognize a fire person even if you spoke face to face with that person . . ."

The essence of the text on this page refers to claiming credit for being smart (". . . You dinks say I'm smart . . ."), avoiding detection by a fire marshal following an arson, and several references ranging from Hitler being a Nazi god to phrases that appear to be biblical references ("Hurite and other illegitimate sons of Joseph") but are probably gibberish in nature. It may be worthwhile to see if a translation of the Native American passage can enhance our understanding of the attempted communication. At the very least these communications characterize the writer as either mentally disorganized or intentionally feigning such, since he says, ". . . I love, I love screwing with their heads with stuff . . ."

This individual acknowledges he has "problems" and points to this in the passage that reads, "Don't you get it? I sent those snap shots to you . . . trying to get you to want me in a sexual way a tenth of the way I'm gonzo over you; ya I'm sick but damn, woman, give me a break, this is overpowering."

In a different passage he says, "I like you in so many ways it's weird especially since I don't even know you, but (if?) the personality profile I pieced together pans out you're a rather remarkable chick."

Summary and Recommendations

1. In summary, the writer's current behavior would be most characteristic of category 3.
2. The analysis of the writer's communication indicates the following levels of concern:
   a. Level of organization of communication: medium
   b. Level of fixation and blame: high for emotional distress, low for physical harm
   c. Level of focus upon a specific victim: high for emotional distress, low for physical harm
   d. Action imperative: medium for emotional distress, low for physical harm
   e. Time imperative: low for emotional distress, low for physical harm
3. The writer's communications do not now indicate a high risk for physical harm. However, the writer's communications are indicative of intent to cause emotional distress to the victim/co-worker even at some limited risk of being identified and potential job loss.
4. Further communications would be very important, as they would minimally indicate that the writer is angry and pressured to have the victim in additional pain and distress and is willing to incur additional risk of identification to accomplish this.
5. Further communications should be carefully analyzed for content that would indicate increased action and/or time imperative, which would increase the risk for physical harm to the victim.
6. The writer does not have a sophisticated knowledge of handwriting analysis as the communication does not show evidence of attempts to alter handwriting. Further, some letters in the handwriting are quite distinctive ("i, y, g") and will lend themselves to comparison analysis with handwriting samples on incident reports, personnel forms, or other such items from potential suspects.
7. Review recommendations and action options with company labor law counsel for input as appropriate.
8. Consult if new information/actions occur.

Risk Control Actions

1. Share profile information with local law enforcement; local individuals may be able to identify him quite easily. Many small towns surround the city and this person will likely stand out to the locals.
2. Treat this individual as a local law enforcement problem, possibly an individual the police would like to question in the aftermath of suspicious fires or if other prominent local females have received letters of the type sent to Ms. Doe.
3. Consider that he may take a "straight" approach, i.e., write as a fan requesting signed picture or autograph. Requests of this sort postmarked anywhere nearby should prompt a closer look by security.

Chapter 9

Threat Assessment Feedback for Management

The threat response team needs clear expectation of the threat assessment professional, whether internal or external, who does the employee interview. These feedbacks are very different from what might be expected. The report should be brief and to the point. All extraneous information should be limited. This tendency to explore side issues that distract from the group's focus on developing risk control options. The feedback needs to be tightly structured to address five points. After receiving this data the group can proceed with the risk option review and the gathering of other information independent of the threat assessment professional needed for organizational decision making. The threat assessors may have accumulated a great deal of data and part of their role is to filter this information down to five key points. Ideally, feedback can be available to the team within thirty minutes of the completion of the interview.

FEEDBACK ELEMENTS

Nature and Context of the Interview

Address information that was provided to the employee as to the purpose of the interview and the conditions under which it will be conducted. Included here are the facts that it is part of a management review process, that the interview is nonconfidential and on the record, and that the interviewers will be sharing the information with the group. The category designation that best fits the situation is provided.

Behaviors/Comments of Concern

At this point, the grid of behaviors and or comments that have led to the review is presented. The employee is given a chance to respond to each of these points of information before they are given to the threat group. The individual's responses for each concern are characterized as admit, deny, or partial admission/denial with a summary of any clarifying information that the interviewee provides.

Perception of Their Roles

Provide information as to how the individuals see their roles in the events that transpired. The degree of responsibility, if any, they see as their own is reflected here. Also, alternative explanations for the role ascribed to them by others are presented here.

Perception of Others' Roles

Provide information into how the individual sees the roles of others in the situation. Clarifying relations as well as any motives for behavior are presented. This is the time to also describe the individual's perception of the impact of the situation on others.

Game Plan

At this point, provide information on the individual's view of how the situation can be resolved. Roadblocks and barriers are included. The individual's role and the role of others in any resolution are presented. A variety of options that any organization might use to resolve such a situation are presented to the individual with his or her response as to how they might manage that option.

CASE EXAMPLE

1. Mike Gelles and Jim Turner of International Assessment Services, Inc., (IAS) interviewed Mr. Jones on December 22, 2001, at the corporate office. Prior to the interview, Mr. Jones was informed by local management of the non-confidential management review nature of the interview and his right to union representation. He was also informed both orally and in writing that the interview was related to concerns that had been raised about work-related behavior, that the company had engaged outside consultants to provide a review, that the inter-

view was on the record and nonconfidential in nature, and that information would be relayed to management. He was informed that he could take a break at any time and that he could consult privately with his union representatives at any time. We reviewed the conditions of the interview and provided him with a copy of "Your interview today." He agreed and we proceeded to interview him for three hours including two breaks. Mr. Jones was responsive to questions during the interview.

2. The following concerns related to Mr. Jones' at-work conduct were discussed with him.

- Corporate employee Tom Doe reported that on January 5, 2001, following a verbal dispute over how to conduct an assignment, Mr. Jones grabbed Mr. Doe and threw him to the floor.

  In response, Mr. Jones acknowledged that he did grab Mr. Doe's belt and pushed him, causing Mr. Doe to fall to the floor. He denied grabbing Mr. Doe by any body part. He further stated that Mr. Doe had "insulted" Mr. Jones' recently deceased father, which preceded his action.

- Corporate employee Maya Ortiz reported that on December 31, 2000, Mr. Jones confronted her. He questioned why she was engaged in a particular activity. She told him that she was working on a specific job assigned by her crew chief. She reported that Mr. Jones told her he did not like the way she was talking to him. He directed her to a meeting with Supervisor Bobby Ball and union representative Robert Smith. During that meeting, she reported that Mr. Jones commented to Mr. Smith, "Come on, Robert, you know it's all about skin," as he was pointing to his skin and added, "You people cover for each other."

  In response to this report, Mr. Jones denied making this statement, indicating that he only stated: "you people" while scratching his hand.

- Employee Mike Roberts reports that he and Mr. Jones disagreed about overtime pay issues. Mr. Roberts stated that Mr. Jones grabbed him and swung him around, causing him to fall to the ground.

  In response to this allegation, Mr. Jones denied touching Mr. Roberts. Mr. Jones stated that Mr. Roberts pushed him with his chest. He identified two other employees (Mel King and Gary Borden) who witnessed Mr. Roberts push him, and stated that these employees would confirm his report of this incident.

- On November 21, 2000, Supervisor Joan Sweet reported that postal employees had observed Mr. Jones using profanity toward postal employee Manny Janner while attempting to tell him to move mail to a specific location in the loading area.

  In response to this allegation, Mr. Jones acknowledged that he had used profanity and "yelled" at the postal employee, Mr. Janner. He stated that he yelled at the employee saying, "f**k this mail" after the postal employee banged into his mail cart with a pallet jack. Mr. Jones stated that he met with Ms. Sweet and apologized to the postal employee. Both Sweeney and Jefferson confirmed this.

- On November 26, 2000, employee John McCallum reported that Mr. Jones confronted him about being early to work and that he should not start working. Mr. McCallum stated it was approved by the supervisor and began to work. At that time, Mr. McCallum reported that Mr. Jones turned to him holding a box cutter with blade extended and said, "You get your mother f*****g hands off my g******n mail." Mr. McCallum then walked away and contacted his supervisor, Mr. George Black, at extension 0245. Mr. McCallum later reported the incident to ABCD supervisor Joan Sweet.

  In response to this allegation, Mr. Jones acknowledged holding a box cutter at the time. However, he denied using profanity and then accused Mr. McCallum of trying to get him in trouble because Mr. Jones had reported Mr. McCallum for sleeping on the job. "He is trying to discriminate against me."

- On December 1, 1998, Supervisor Jay Brad reported that while discussing rules for overtime pay Mr. Jones became angry, raised his voice, threw his company security badge on the desk, and walked out of the office. This was not reported at the time.

  In response to the allegation, Mr. Jones acknowledged that he listened to Mr. Brad, was upset that his hard work was not appreciated, placed his badge on the desk, and then walked out. He denied he raised his voice and threw his security badge on the desk.

- On December 18, 2000, employee Ti Mau reported trying to load some mail on a cart to which Mr. Jones was attempting to hook his tractor. Mr. Mau stated that he asked Mr. Jones to stop. Mr. Jones began to yell at him (inaudible) and then got closer to him and stated, "I'll kill you, you mother f*****." Mr. Mau stated he looked him straight in the eye and said "Mister, you are not going to do that. I'm not going to take your crap anymore. I am going to call the police." Jerry Lightfoot, who was working with Mr. Mau, witnessed the interaction and stated that he heard Mr. Jones say the word "kill" and then heard Mr. Jones say, "Go ahead and call the police. I am going to kill your ass."

  In response to this allegation, Mr. Jones denied ever making a statement that included kill, but he did not recall what he actually said. He stated that Mr. Mau approached him yelling at him (no specifics) and that "he bumped his head into my hat."

- Results of interviews from employees indicate that Mr. Jones uses profanity, yells, and gets "up into other people's faces." All of the employees interviewed stated that they were distressed by Mr. Jones' behavior in the workplace.

3. With regard to the employee's view of his present situation and the reports of his behavior, Mr. Jones denied or modified by reduction many of the allegations against him. He acknowledged that he "does have a temper" and needs to work on controlling his temper. He reported that significant people in his life had told him he was going to get in real trouble if he didn't learn to control his temper. He was unable to articulate how he would work on his temper or what he needed

to change. He repeatedly stated that he was "a hard worker, made money for the company, had perfect work attendance, and no one worked as hard" as he did. Mr. Jones repeatedly stated that "others discriminated" against him, but could not provide specifics of any incident. When asking Mr. Jones for specifics regarding his behavior, he repeatedly stated that he was telling the truth and "only God could justify what was right and wrong."

4. He stated that he felt Mr. Mau and Mr. Lightfoot were lying and that they made up the story to "discriminate" against him because Jones had reported Lightfoot for sleeping on the job. He stated that people misunderstood his behavior and needed to learn to just leave him alone.

5. When asked about how he might respond to the current situation, Mr. Jones did not take any responsibility for his behavior aside from his grabbing Mr. Doe "by the belt." He stated that the allegations made against him were the result of being discriminated against by his co-workers. He was unable to view the situation from a management perspective (e.g., management being faced with a series of different employees and nonemployees making similar reports over time). He responded by continuing to state that he was telling the truth. Although he did identify his temper flare-ups as problem behavior, he could not formulate any specific plan to resolve this problem other than for others to leave him alone. When asked how others might do their jobs while leaving him alone, he was unable to give any ideas.

With regard to his response to the various options that the company might employ in response to his behavior, Mr. Jones denied any wishes to harm anyone and denied owning any weapons. He stated that if he were to be terminated or disciplined, he would retain an attorney and fight the action.

At the end of the interview he stated that he had been treated respectfully. His union representative was given the opportunity to make a statement and you have been given that brief statement.

Chapter 10

Stalking As a Contemporary Crime in the Workplace

BACKGROUND

Although the stalking of others with the intent to cause distress and/or harm has appeared in history and literature through the ages, Western society's concern with stalking is relatively new. Many employees and employers struggled to cope with the problem with limited knowledge or resources for assistance. In this chapter we shall not attempt a comprehensive treatment of stalking as several volumes do that (Meloy, 2000; Davis, 2001b). We shall focus on responding to the issue in an organization's environment, specifically, how the information enters the environment and risk control options.

History of Stalking

In the late 1950s, Beverly Hannon was working as a secretary in a Midwest office. Her office window overlooked the sidewalk. One day, a Korean War veteran walked by her window, saw her nameplate on her desk, and came to believe he had been engaged to her at one time. He also believed that she had written him a Dear John letter while he was in the Korean War. His cards, letters, and phone calls to Hannon became overwhelming. He began to follow her everywhere. Several months later, Hannon was working late at the office when her stalker slipped through the office front entrance and found her at her desk. She was able to evade him on this occasion. The stalking continued for ten years, and it was not until twenty years later that Hannon reported truly knowing it was over when she read the stalker's obituary.

The pivotal point in contemporary history was the 1989 stalking of Rebecca Schaeffer, a twenty-one-year-old Los Angeles television star, by Robert Bardo. Bardo had been romantically obsessed with other entertainment industry stars such as Olivia Newton-John and Debbie Gibson. In his stalking, he discovered the security where they worked to be too good. He then fixated on Rebecca Schaeffer. Bardo tried to get to Schaeffer twice at her workplace, Warner Bros. Studios. Security personnel turned him away. He then found out where she lived by hiring a private detective to obtain her address from the California Department of Motor Vehicles files. Bardo went to her home. He was initially surprised when Schaeffer, working at home and expecting the delivery of a new script for review, answered the door herself. Bardo returned a short time later. He grabbed Schaeffer as she once again opened the door, and killed her with a .357 Magnum handgun. Bardo was arrested, and spoke openly of his stalking and eventual murder of Schaeffer.

Although there had been other equally tragic outcomes to stalking, this case captured the attention first of California, and then of the rest of America. As a consequence, California became the first state to pass an antistalking law (SB 2184, which became California Penal Code 646.9) in 1991 and amended in 1994. The rest of the states and the District of Columbia followed. For a review of the legal issues, refer to the excellent chapter by Rhonda Saunders (1998).

The Schaeffer stalking and homicide was also instrumental in the 1990 creation of the first specialized investigative unit for stalking crimes, the Threat Management Unit (TMU) of the Los Angeles Police Department. Directed by Lieutenant John Lane, this unit consists of a commanding officer and detectives who investigate and arrest stalkers. The unit has received international recognition for its leadership in assisting individuals and employers faced with this problem. Intrigued by this work, other law enforcement agencies around the country have become more aware of the need to address and intervene early in stalking crimes. This has led to the formation of the Association of Threat Assessment Professionals (ATAP) with membership limited to law enforcement officers, private security professionals, and mental health professionals involved in this area of public service and investigation. Enhanced education and professionalism is changing the focus of both law enforcement personnel and district attorneys to increased early intervention in stalking cases.

Two significant developments in the 1995-1996 time period hall-marked the growing movement to treat the crime of stalking more seriously. First, a jury in Los Angeles deliberated for only five hours before finding Robert Dewey Hoskins, age thirty-eight, guilty of stalking recording and film star Madonna (Saunders, 1998). Hoskins had twice gone over the walls of Madonna's Hollywood home and threatened to kill her if she did not consent to be his wife. Hoskins was sentenced to ten years in prison. The message from the jury was clear: the crime of stalking was going to be taken seriously.

Second, at the direction of President Clinton, a twenty-four-hour toll-free hot line was established for victims and potential victims of stalking and domestic abuse. The number to call for information and assistance is 800-799-SAFE.

Victims of Stalking

A popular myth is that victims of stalkers are primarily individuals with high public visibility in politics, the entertainment industry, and the media. In fact, data from the Threat Management Unit of LAPD indicate that over half of their cases involved ordinary citizens (Zona, Palarea, and Lane, 1998). Further, while the movie *Fatal Attraction* popularized the idea of the female stalker of the male victim, the majority of stalkers are male and the majority of victims are female (Tjaden, 1997). Most individuals will never have any contact with a stalker. However, for those who do become stalking victims, the experience is both unreal and embarrassing. Consequently, most victims have difficulty severing all contact immediately. The unfortunate result is that the stalker feels reinforced by any contact: positive, neutral, or negative. Characteristically, when the employee informs the employer of this behavior, the employee is likely to have already engaged in a series of unsuccessful coping behaviors that have been interpreted by the stalker as reinforcement.

The Stalkers

As noted previously, the majority of stalkers are male. However, sufficient cases of female stalkers are documented for research to occur (Purcell, Pathe, and Mullen, 2001). The stalker generally knows the victim through a domestic or personal relationship. The stalker is generally intelligent and can be very elaborate in preparations for contact with the victim. When confronted, the stalker may initially deny the behavior, and then attempt to convince others of rational explanations for the behavior. The stalker may overtly indicate an understanding that this behavior is going to progressively lead to trouble with the legal system, yet will reinitiate the behavior within days of such a statement. Although the stalker may continue to successfully maintain a job and other personal relationships at the same time as the stalking behavior, he or she may be willing to sacrifice job or personal relationships as necessary to continue the stalking behavior.

The subject of stalker typologies has been examined in a number of articles (Boon and Sheridan, 2001; Holmes, 1993; Zona, Sharma, and Lane, 1993). These systems will be of interest and use to mental health professionals, but we urge extreme caution with their use in the organizational setting. Some terms are usually not fully understood by those outside threat assessment and may lead threat teams within the organization to begin to label individuals. Subsequent review of decision making may focus on whether the right label was used by the group rather than on the actions taken by the group. The category system outlined previously is designed to keep individuals focused on the nature of the risk and responding to control that risk.

Stalking and the Workplace

Stalking can be a type of workplace harassment and violence. Stalking presents itself in the workplace through the behavior of: (1) one employee against another employee, (2) a nonemployee (current or former romantic partner) against an employee, or (3) a nonemployee (little or no prior contact) against another employee. Such stalking behavior is not characteristically confined to the workplace, but will involve off-work acts against the employee as well. In some cases, as a result of emergency housing relocation, the workplace may become the only place where the stalker knows to look for the individual.

Managers of organizations do not need to become fully knowledgeable about all of the management and criminal law issues related to stalking; however, an overview of these issues will assist in working with external threat assessment consultants and external labor law counsel. Organizations are best not diverted into determining if the behavior meets the legal criteria of stalking, which is the realm of legal and law enforcement. The threat group needs to focus on the impact of the behavior, whether the behavior violates company rules or procedures, and how to respond to the possible increased risk. In cases where a more in-depth knowledge is required, see Meloy (2000) and Davis (2001b).

STALKING DEFINED

Stalking has generally been defined as having three components:

1. a course of conduct in which there is a repeated pattern of following or harassing another person,
2. making a threat to harm or acting in a threatening manner, and
3. intent to cause harm or distress.

Stalking Component #1: Course of Conduct

The first component of stalking is course of conduct, which means that the threats or acts of harassment need to occur more than once in most circumstances. There must be a series of such incidents generally linked to a continuity of purpose to cause distress in another person. To understand this, it is necessary to view stalking as occurring across three variables:

1. frequency of stalking acts (high versus low),
2. the type of stalking acts (e.g., harassing versus following behaviors), and
3. the point in which the stalking acts become a law enforcement matter.

Frequency of Stalking Acts

Let's examine the variable of high versus low frequency of stalking behaviors first. On one end of this variable, for example, a male employee becomes romantically obsessed with a female employee. He provides unsolicited gifts to the employee at work, appears uninvited at her residence, and repeatedly follows her in his car at the end of the workday. This series of events would very likely meet the course of conduct criteria. On the other end of this variable, a single unanticipated or unwanted appearance of that male employee at the other employee's home, or a hostile encounter in the company parking lot by competitors for the attention of the employee, would not meet the course of conduct criteria. The course of conduct issue becomes less clear when we look at the time frame within which the stalking behaviors need to occur. If several stalking behaviors take place within a single day, would this meet the course of conduct criteria? State laws vary on this issue and legal review by the district attorney's office in the local jurisdiction where the acts occurred is necessary to clarify such issues.

Type of Stalking Act

The second variable is the type of stalking act. Course of conduct has been described not only in terms of frequency of acts but by the nature of the acts themselves, such as whether the victim is being harassed and/or followed. The exact definition of what behaviors constitute stalking varies from state to state as well as by the Federal Interstate Stalking Punishment and Prevention Act of 1996 (Title 18 USC Section 2261). Again the threat response group's job is not to determine whether a law has been broken but whether the conduct is disrupting the workplace and/or apparently violates company policy. A determination needs to be made whether a potential threat triggers the threat assessment process and risk control review. If information is available that a law has been or is likely to be broken, this additional information should be considered in the threat assessment and risk control deliberations. Legal issues need to be determined by the professional educated and licensed to do so. Team members need to exercise care in rendering legal or law enforcement opinions when this is not their area of expertise.

Some behaviors that stalking may include are:

- Following or appearing within the sight of an individual
- Approaching or confronting an individual in a public place or on private property

- Entering, remaining in, or delivering items to property owned, leased, or occupied by an individual
- Contacting an individual by telephone or fax
- Sending mail or electronic communications

Some states accept that stalking may also include threats and acts of harassment to family members of the victim (California Penal Code 646.9). In California, the acts of stalking are defined not by a list of prohibited behaviors but rather as broad behaviors that reasonably cause fear and distress in the victim.

When Stalking Becomes a Criminal Violation

Finally, in time in which stalking acts become a criminal violation may vary. For example, Texas begins to count the stalking acts after the victim has reported any one act to law enforcement (Texas Penal Code Annotated 42.07; Texas Code of Criminal Procedure Annotated 17.46, 42.12, 42.8). Stalking acts prior to this first report to law enforcement, even if numerous, are not eligible to show course of conduct. North Carolina requires that the victim or a third party designated by the victim inform the perpetrator that the stalking behavior cease before the state stalking law can be invoked (NC General Statutes 14-277.3).

Stalking Component #2: Direct Threats to Harm and Threatening Behavior

Stalkers are frequently intelligent, adaptive, and creative. They display a range of behaviors to their victims. There may be explicit direct verbal threats of intent to cause physical harm. There may be indirect threats of harm occurring to the victim by unknown persons or groups if the stalker's needs are not met by the victim. Threatening communications may range from drawings or dead flowers to e-mail and online contact. Nonverbal behaviors such as a sudden appearance in unexpected or uninvited situations, motioning, and suggestive body movements may occur. For employees, stalkers will typically communicate through telephone, U.S. mail, interoffice mail messages, e-mail, and gift deliveries. The stalker may self-identify in some of the communications, and not in others. The style and content of the communications may be so similar as to functionally identify the stalker. Communications containing a combination of limited words and drawings or cutout pictures are common. Drawings or pictures often contain images meant to resemble the victim accompanied by images of knives, guns, and/or blood.

Online stalking has begun to appear on electronic computer bulletin board systems. A Fresno, California, resident and teacher using the name Vito used a commonly used system to target in New York City. Escalating from the initial computer message insults and hate mail, Vito sent a computer fax to the individual's employer, stating the employee was a convicted rapist and molester of children. Vito also sent e-mail messages through computer systems stating that the individual was HIV positive. Vito targeted a number of other individuals as well, resulting in his eventual apprehension. Such cases covering multiple jurisdictions and application of federal law are complex and require the consultation of experienced legal experts in this specialized area.

Many state antistalking laws further require that the verbal statements or behavior by the stalker constitute a credible threat. This issue of credible threat can be a difficult point if the stalker has a lengthy history of mental illness, and is making threats in a highly disorganized manner to multiple victims. In some states, law enforcement agencies may not view this as a credible threat and may not be supportive of an arrest under an antistalking statute. However, this individual's behavior should not be immediately discounted and ignored, as it may qualify for an arrest under other state penal code sections, such as terrorist threats or involuntary psychiatric hospitalization as dangerous to others. Regardless of whether the legal system chose to act, victims still experience feelings of being threatened and distressed by the acts. These need to be addressed by the organization and lead to the activation of the threat assessment process.

Stalking Component #3: Intent

Issues of intent are consistently reflected in both the psychology of crime and in criminal law. If two employees become involved in a physical fight on the building's loading dock and one employee is severely injured, such severe injury may have been intentional or may have been the result of the employee falling down during the fight. Further, the uninjured employee may have had a mental disease or defect that prevented him from appreciating the wrongfulness of the acts at the time. Consequently, state antistalking laws define intent by the use of words such as willful, malicious, knowing, and purposeful. Issues of intent may become important in a stalker's postarrest defense in that the delusional stalker who is love obsessed with his victim may assert that all of his behaviors were designed to win his victim's attention and affection, not to harm her. To address this potentially problematic issue of intent, states have made efforts to move away from proof of intent, requiring rather the willful course of conduct that would reasonably cause fear, terror, and distress, and, in fact, does produce those effects in the victim.

STALKING AS A MENTAL DISORDER IN THE WORKPLACE

Not all stalkers have a diagnosable mental condition or illness, but some stalkers have been found to have mental disorders such as:

- *Delusional Disorders:* Usually present in normal-appearing individuals who strongly hold to a belief system for which no objective evidence exists.
- *Manic Depressive Disorder:* A disorder in emotions and behavior can be very accelerated and pressured, often resulting in a manic pattern of grandiose plans and beliefs about personal relationships with others.
- *Personality Disorders:* Seen in individuals with chronically maladaptive ways of dealing with the world in general, which may also include attempts to live out a fantasy about a relationship with someone else.

The stalker may have a delusional disorder called erotomania. The stalker has come to believe that the object of his or her love, a powerful or popular person, is truly meant for him or her, even though this person may have never met the stalker. Further, the stalker believes that the object of love is passing secret messages intended only for him or her. The stalker is also convinced that if they indeed did meet under the right circumstances, then all would be perfect forever. The historical basis for this diagnosis comes from the work in the 1920s of a French physician named Clerambault, whose patients included a fifty-three-year-old Frenchwoman obsessed with England's King George V. This patient saw all British tourists as agents of communication to her by the king, and believed King George was communicating to her from behind moving curtains in Buckingham Palace (De Clerambault, 1942). If the stalker with erotomania is identified and required to seek psychological treatment, the potential for ceasing the behavior is limited. Repeated arrests and court-ordered psychiatric hospitalization represent a common approach that may manage the risk, but not eliminate the erotomaniac's stalking behavior.

Other stalkers may have manic depressive (bipolar) disorder. In stalkers, manic or intense, highly pressured activity is present. A number of medications are available to treat this disorder including lithium. As a result, this type of stalker has been primarily responded to by the mental health system rather than by the criminal justice system. Unfortunately, a number of stalkers with manic depressive (bipolar) illness find that life with the proper medication levels is less interesting than being in the somewhat exciting manic phase. As a consequence, the stalker may discontinue medication and once again become a problem.

Stalkers may also have a personality disorder that may include a belief in a right to have access to things or to people just because they have a fantasy need for such access. Although many criminals of all types have personality disorders, this has not been accepted as a successful defense against criminal or personal liability. Stalkers with personality disorders have been primarily responded to by the criminal justice system rather than by the mental health system.

In these cases threat teams need a general background on the nature and types of stalking and stalkers. Threat response groups are not in the business of diagnosing and should refrain from such activities. Again the focus needs to be on a thorough threat assessment, identification of risk factors, and the establishment of risk controls. The use of a psychologist or psychiatrist with a background in threat assessment is crucial in such cases. These individuals, however, are not in the business of trying to treat the stalker. Their role is to provide probable behavior patterns and risk control actions. As well as providing experienced opinions on the outcomes of certain types of risk control action, the individual chosen to serve in this role needs to have extensive experience in managing threat behavior in organizational frameworks.

THREAT ASSESSMENT AND RESPONSE

A series of issues may arise in workplace stalking situations. Due to the nature of cases involving questions of romance, strained marriages, or intimate relationships, sensitivity is required. The challenge facing the threat management team is balancing safety and privacy. The first question that faces the organization is how the information has entered the system and how to begin the initial response.

The information will enter the system from four primary channels:

- *Situation 1:* Employee reports stalking outside of work; no stalking at work. The information is provided to the organization from the employee about being stalked outside of work and a fear of the stalker coming to the workplace. This often occurs simultaneously with the employee personally making a police report and/or obtaining a restraining order.
- *Situation 2:* Employee reports receiving threatening communications at work. Information is provided to the organization from the employee as a result of receiving threatening communications at the workplace. Such communication may or may not be known to co-workers.
- *Situation 3:* Co-workers receive threatening communications at work, unreported by target. The information is provided to the organization from co-workers who receive or intercept threatening communications directed to the employee. This interception occurs accidentally by co-workers or may be intended by the stalker to be seen by co-workers.
- *Situation 4:* Co-workers are fearful of unreported stalking situation. The information is provided to the employer from co-workers who have learned of the as yet unreported stalking situation usually from the target or from friends/relatives. The co-workers have become fearful for their own safety.

With situation 1 (stalking outside of work; no stalking at work), the threat assessment team needs to consider the following options:

1. Interview the employee to further determine the nature of stalking behavior, e.g., type, place, frequency, means.
2. Interview the employee to determine what evidence or behavior has led the employee to be concerned that the stalker may come to the workplace.
3. Educate the employee about the employer's limited responsibility for maintaining a safe workplace versus the individual's broader responsibility for personal safety outside of the workplace.
4. Determine if the employee wishes to have a referral to the company employee assistance program.
5. Caution the employee that the safety of all employees at the work site may require the involvement of corporate security, law enforcement, or external consultants in law and psychology.
6. Contact relevant law enforcement agencies to determine their assessment of the current situation and obtain copies of police reports and other available court documentation.
7. Establish a procedure for recognition and response to the alleged stalker if appearance at the workplace should occur.

With situation 2 (employee receives threatening communications at work), the employer may wish to consider the following options in addition to the options previously described for situation 1:

1. Maintain the chain of evidence of the communications received and avoid contamination of evidence by additional fingerprints, etc.
2. Contact relevant local law enforcement agencies and offer to provide evidence received.
3. Consider the development of a strategy to intercept additional communications directed toward the employee/victim.
4. Evaluate the threatening communications through consultants.
5. Make risk assessment based upon threatening communications received and any other information known about other stalking acts.
6. If stalker can be identified, consider need for intervention by law enforcement contact, restraining order, or disengagement strategy.

7. If stalker cannot be identified, consider use of documents expert or voice print expert to establish if same person is likely responsible for all threatening communications.
8. Evaluate possibility that employee may have authored threatening communications for his or her own reason.

With situation 3 (co-workers receive threatening communications at work), the employer may wish to consider the following options in addition to the options previously described for situations 1 and 2:

1. Consider whether co-workers have accidentally received threatening communications intended for another employee or whether the stalker has now decided to involve co-workers in the harassing behavior.
2. Evaluate the probable intent of the stalker and the emotional impact upon the co-workers who are now also involved in the situation.
3. Actively pursue the development of a strategy to intercept additional such communications directed toward the employee/ victim and receivable by co-workers.

With situation 4 (co-workers are fearful of unreported stalking situation), the employer may wish to consider the following options in addition to the options previously described for situations 1, 2, and 3:

1. Caution reporting co-workers that the safety of all employees at the work site may require the involvement of corporate security, law enforcement, or external consultants in law and psychology.
2. Determine if reporting co-workers wish a referral to the company employee assistance program.
3. Develop strategy for interviewing alleged employee target.
4. Interview alleged employee target with attention to issues of privacy balanced against need for a safe workplace.
5. If stalking behavior is confirmed and employee is cooperative, solicit additional information, and involve law enforcement and external consultants in psychology and labor law.
6. If employee is uncooperative, reinterview reporting co-workers for additional information and assess need for direct contact with

law enforcement and enhancement of existing security procedures.

7. Consider use of external consultants in psychology and threat assessment for additional interview of initially uncooperative employee.

When referrals to the employee assistance program are made, a member of the team should contact a manager at the employee assistance program and explain that the organization has a situation, which it is managing, and employees may be contacting them for assistance. Also, provide a contact number and name with the team. This will alleviate the employee assistance program's staff from becoming alarmed that an uncontrolled situation exists as a series of employees call for assistance.

Stalking behavior in the workplace comes from four primary sources: organizational employees, organizational clients, organizational vendors, and individuals unrelated to the organization. In addition to the previously outlined actions, some other items need to be considered in each of these situations.

1. Behavior by organizational employee:
   - Review by human resources/security to see if a pattern of behavior is emerging
   - Interview supervisor to determine change in work requirements dictating interaction
   - Review job situation requirements
2. Behavior by organizational client:
   - Review past difficulties with client; may require interviewing individuals in organization with knowledge of prior unreported problems
   - Consider arranging observational opportunity for management of work interactions, if possible
   - Consider with management assignment of client to other individuals; Recognize this often is more complex than it may seem as a matrix of customer satisfaction, financial incentives, and job performance concerns may arise
3. Behavior by organizational vendor:
   - Review with materials management contractual issues

- Review with legal counsel impact of involving vendor organization in the review process
4. Behavior by individual unrelated to organizational mission:
   - Review with security any pattern of behavior
   - Review security and legal options to include a no-trespass notice and/or restraining order
   - Request security liaison with local law enforcement

In relation to the recipients of such behavior, the threat assessment team needs to ensure that three other concerns are addressed: boundaries, monitoring, and security briefing.

Boundaries

In many cases the first contact with a victim in an investigation may lead to providing him or her direction on establishing clear boundaries. In some cases it may be directing the victim to first offer a cease and desist communication or for the investigator to take that action on behalf of the victim.

In cases where the subjects remain fixated and continue to escalate, they frequently violate established boundaries. This can range from ignoring a cease and desist communication or directly violating a restraining order or condition of probation or parole.

In any case, the victim should have no contact with the subject, unless directed to do so by law enforcement. Although this sounds like common sense, it is common to learn that a victim has communicated with a subject and has thus reinforced his or her pattern of behavior, minimizing any earlier attempt to establish a boundary.

For example, a former employee began writing and visiting his former supervisors at their homes in another state demanding to know why he was fired ten years ago. Some time had passed when the five former supervisors began receiving letters from the subject regarding their "homosexual love" for him. None of the supervisors were reportedly homosexual and were very angry and disturbed by the subject's statements. They were so disturbed, in fact, that they felt compelled to respond to his comments, even after they had previously agreed not to, and had no problem being unresponsive in the past. This renewed contact with the subject escalated the situation and made it necessary to serve the subject with an arrest warrant. The subject then refocused onto federal and state law enforcement officers as interfering third parties.

Monitoring

The victim should be encouraged to engage in continued monitoring and be sensitized to potential boundary probes and indicators to potential escalation. It is helpful to suggest that the victim start a contact log that includes date, time, background noises, tone, context, and statements (verbatim). Law enforcement may assist the target in developing a safety contingency plan. This plan should include a list of safe places and critical telephone numbers. The plan will also include other recommendations such as keeping a full tank of gas, a backup set of car and house keys (with trusted neighbor), a packed suitcase (for a quick departure), access to reserve money, medication, passports, social security cards, and birth certificates (Office of Victims of Crime, 2002).

Last, the victim should be educated in regard to the timeliness of notifying, alerting, and reporting to the threat management team via security or human resources new information pertaining to the situation. Each additional contact or attempted contact needs to be evaluated for signs of potential violence escalation. Once the threat assessment team obtains the relevant information, a discernible pattern may emerge that will greatly assist in making accurate and useful threat assessments and reviewing control actions.

Security Briefing

The victims need to be provided a reasonable security briefing, preferably by law enforcement, that covers basic security measures at home, work, and in a vehicle, as well as personal safety options they may wish to consider. An effective law enforcement liaison via security or management will enhance the likelihood of this happening. These briefings may also include the pursuit of protective actions through the courts that include restraining orders and no trespass orders.

Chapter 11

Cyberthreats

Although we have focused considerable attention on communicated threats and behaviors that are observable in the public domain, some attention should be given to the factors associated with the increase in cyberthreats and cyberstalking.

DEFINING CYBERTHREATS

The age of electronic communication has expanded the manner in which individuals communicate, develop relationships, and intentionally inflict emotional distress. Cyberthreats, cyberstalking, and cybertrolling are all terms for which very few studies and literature offer definitions. A limited amount of published information provides guidance to the threat management team on managing cyberthreats.

The increase in the use and availability of the Internet, as well as its low cost and ease of use, has created a readily accessible medium for former employees, customers, and anonymous individuals to engage in threatening communication, harassment, and stalking. Although there is no universally accepted definition of cyberstalking, the term is used in this chapter to refer to the use of Internet, e-mail, and other electronic communication devices to stalk another person (Office of Justice Program, 2001).

Stalking, as discussed in Chapter 10, involves repeated harassing and threatening behavior such as following a person, appearing at a person's home or business, placing harassing phone calls, leaving written messages or objects, and vandalizing property (Office of Justice Program, 2001).

Although the communication via electronic means is similar in the end result, in that it causes considerable emotional distress, fear, and terror for the victim, cyberstalking provides a different opportunity for individuals who want to remain anonymous. It also widens the range of individuals who may be tempted to engage in threatening and harassing behavior. Such individuals would be unlikely to write directly via traditional means or to orally communicate a threat, pursue, or stalk an individual. Individuals who communicate threats via the Internet may or may not move along a continuum to more formal stalking. They also may not attempt to exert control over their victims offline.

Anonymity often serves the cyberstalkers' sense of security and safety as it relates to being identified. The loss of external constraints in the cyberworld empowers the individuals to take some action that they might not do in person. In cyberspace, behavior is not scrutinized or observed. This increases the ease with which an individual might transgress constraints that exist in the public world. The result is a less inhibited impulse that translates into disturbing communications and behavior online. In the public world, rules and social norms, observable by others, restrain an individual's behavior. Society collectively holds individuals accountable and applies consequences to behavior. In the cyberworld individuals perceive that they are bound only by their own internal constraints. In cyberspace they exist in a world where one is only accountable to oneself.

In cyberspace communicated threats, harassing investigations, and violation of privacy can be conducted anonymously. This form of harassment may have become more popular for stalking behavior because law enforcement is associated with traditional threats and harassment made in the public world. The difficulty in identifying the author and the lack of physical evidence make the Internet a much more attractive medium to communicate and discharge frustration, disappointment, and entitlement. With several clicks of a mouse the cyberstalker can locate personal information that can be clandestinely used to rob individuals' identities, harass them, and stalk them. Cyberstalkers can access enormous amounts of personal information (Office of Justice Program, 2001).

RISK CONTROL

Risk control begins with monitoring the types, accuracy, and quantity of information available about the company, employees, and practices on the Internet. This includes communications about products and services. Searches can be conducted by individuals at the direction of the team to determine if company-related sites are a possible information source.

Similar to communicated threats in writing, orally in person, or on the telephone, communicated threats via the Internet are evaluated along the same criteria. The statements are assessed in regard to organization, fixation, focus, action, and time imperative. The communication needs to be saved and analyzed for address and time stamping. Instead of a postmark the time and date are noted by the Internet service provider or server. Identification of the Internet service provider can lead to options related to service.

Management strategies need to include monitoring the e-mail box for continued communication much the same way a telephone voice mail is monitored. If the author is known, the appropriate boundaries need to be articulated. Often a cease and desist communication needs to be directed to the author, and electronic mail harassment charges should be filed based on local statutes. Laws surrounding telephone harassment, communicated threats, and stalking exist, but setting clear boundaries remains critical to managing the subject. In cases where cyberspace communications are made anonymously, liaison with law enforcement and the use of computer forensics experts can be useful in identifying the subject. Without computer forensics and the participation of law enforcement, it is difficult to do anything more than monitor. Setting up separate mailboxes for the target individual will remove some of the disturbing effects of the harassment while allowing the threat management team to monitor communications until a law has been violated or the author has been identified. Although a number of states have passed laws against electronic harassment, a debate remains about whether electronic threats and stalking constitute criminal behavior or whether the perpetrator is entitled to First Amendment protection (Deirmenjian, 1999).

CASE EXAMPLES

A manager continues to receive lewd and vulgar e-mails. Although the e-mails are anonymous, the author is emotional and includes details about the manager and the workplace that the manager and security personnel believe only an employee or former employee could know. The e-mails appear to have started after an employment action and although not directly threatening the manager, they cause considerable distress. The e-mails also contain information that suggests that the author has obtained personal information about the manager from the Internet and the company Web page.

The threat management team has the communications reviewed for the potential for violence. They also have reviewed the e-mails for details that give some insight into where the person may have worked in the company based on the familiarity with the manager who is receiving the e-mails. The manager is given a new e-mail address while security monitors the previous mailbox for continued communications. The manager does not receive any new communications in the new mailbox but continues to receive similar messages in the old box. This supports the hypothesis that the employee is a former employee. Based on the time and the content of the communications, the former employee pool is narrowed. A communication is sent to cease and desist by corporate counsel. The communications stop.

An employee becomes the target of a cyberstalker as a result of communicating in chat rooms while on the company computer and e-mail account. The online romance blossoms and, based on the address of the employee, the stalker now has the identity of the company. The stalker forecasts that he will attempt to visit her at work and arrives at the workplace asking for her. (This example highlights the need for companies to have Internet use policies in place and to control access to Internet sites that may leave the company vulnerable to a stalking incident on the premises.)

A customer of a company has become very discouraged by the services rendered to him. He designs a Web site that hosts a variety of negative statements about the company. His actions, although not directly threatening, communicate a negative view of the company. Over time the former customer escalates the content of the information posted on the Web and begins to identify specific individuals. The threat management team monitors the Web site for escalation and potential for violence. The team notices that although the content of the communication remains the same, the language is escalating to emotional discharges and the frequency with which information is being added and changed is increasing significantly. The increase in frequency and interest is a more concerning variable associated with escalation than the content itself. A field interview with law enforcement is arranged to assess the degree to which the former customer may pose a threat to the company and its employees.

Chapter 12

Insider Threat:
Risk Management Consideration

Considerable investment has been made by the U.S. government to study the behavioral risk indicators associated with "insider threat" and espionage. In the early 1980s, we learned that the risk to classified information was associated with insiders, and not necessarily secret agents from foreign governments. The results of numerous research projects and criminal investigations determined that government employees and military personnel, who were granted access, were volunteering to share classified information with foreign governments.

Today, there is a very similar parallel in corporate America. At times, competition is fierce among organizations. This has led to concerns about the inappropriate sharing of proprietary information. With changing economic times, organizations are downsizing, rightsizing, and restructuring. The U.S. workforce has changed its attitude toward work and career goals. Long gone are the days when an employee would join a company, planning to spend the next twenty to thirty years there and then retire. Employees are more likely to think about their careers in five-year blocks. Employees are frequently looking for opportunities and better salaries within their careers, not necessarily within a single organization. Corporate executives and managers are increasingly concerned about employees who are referred to as insiders, sharing proprietary information for profit. The insider threat is similar in attitude to the employee who may be a workplace violence threat.

The insider threat has a set of behaviors that arise out of entitlement and dissatisfaction. This chapter will highlight themes learned in studying espionage to articulate the risk of the insider to share proprietary information.

As a result of numerous espionage arrests in the 1980s, the government funded a research project that was initiated as an intelligence community project comprised of behavioral, security, and counterintelligence professionals. The project's goal was to develop an understanding of the behavior, motivation, personality, mind-set, and triggering events of the trusted insiders turned spies. A team of federal agents, government psychologists, and psychiatrists interviewed individuals who had been arrested and convicted of espionage. The interviews focused on the subjects' motivation, awareness of security policy, the triggers, and means that led to committing their crimes. The project generated data that was formulated into publications, training seminars, and policies. Today, the project has evolved into a group of multidisciplinary professionals who continue to study the "trusted insider." This individual of the new millennium is willing to break security procedures and policy, and commit egregious violations through the abuse of information systems, proprietary information, and fraudulent acts.

The threat management team consisting of people with security, human resource, legal, and behavioral expertise can support management when someone may be revealing proprietary secrets to competitors or foreign governments.

BEHAVIOR, PERSONALITY, AND MOTIVATION OF THE EMPLOYEE AT RISK

"At risk" is defined as the possibility that a "trusted insider," who has been granted access to sensitive information, would commit a variety of offenses, from minor policy violations to economic espionage. This behavior typically is not committed as a result of an impulse, but consciously pursued over an extended period of time. Therefore, from a management and security perspective, identifying "at-risk behavior" or "patterns of at-risk behavior," can result in preventing considerable loss of profit and/or proprietary information.

Economic espionage is not the result of a single motive. It is the end result of a complex set of problems, conflicts, and disputes. Generally, espionage results from a culmination of problems that are often reflected in the individual's personal life. A pattern of behavior often leads to frustration, disappointment, and a sense of inadequacy. Generally, individuals who violate trust and commit espionage or egregious violations pertaining to sharing confidential information feel their employer was unresponsive to their needs. In most cases, the individuals' frustration and dissatisfaction conflicted with a self-aggrandized view of their abilities and achievements, which promoted an increased sense of entitlement. This seems to evolve into a very egocentric view of what they believe the organization is or is not doing for them, what the organization owes them, and what they perceive to be their right to acquire. This leads employees to act in ways that result in more immediate gratification and satisfaction, despite what the rules or the laws state.

Often, if their needs are not met or they are not treated in a manner that they feel they deserve, they may act rebellious, exploitative, passive-aggressive, or destructive toward those who they think are neglecting them, or not recognizing and validating their potential. This process can be most evident in behavior that demonstrates their intolerance of criticism, inability to assume responsibility for their actions, blaming others, and minimizing their mistakes or faults.

None of the trusted insiders who have committed espionage entered the government with the intent to violate a trust or commit espionage. In addition, none of the trusted insiders studied in the government project, or reviewed elsewhere in open or classified debriefings, ever suddenly and without preconception committed espionage. Rather, they engaged in a pattern of behavior that reflected a movement psychologically from idea to action. In almost every case, they engaged in behavior that would be termed "at risk" but was not recognized or reported at the time by co-workers and supervisors. These types of behaviors are routinely referred to in ethics and security awareness training and include their responsible handling of sensitive and proprietary information, disclosure or dissemination of information to noncompany persons.

One of the most frequently offered rationalizations by convicted trusted insiders was that "physical security was lax; tighter security would have been more of a deterrent." Whether this was just another attempt to project responsibility onto the organization or an after-the-fact comment, it makes good sense. Physical and information systems security that exists openly in the corporate environment has proven to be a valuable psychological deterrent. Similar to a magnetometer, if a person is considering rule violations and is aware of the potential to be detected, behavior that may prove to be damaging to the company is deterred.

Finally, given the tendency for trusted insiders to disclose their actions to others, it is critical to put in place a reporting mechanism that facilitates the flow of information to management. As with other threats the workforce should be engaged as an active monitor of inappropriate behavior. Policies and confidence in the organization's ability to respond to reported information in a responsible way is a major outcome of a well-functioning threat management team. Often the trusted insiders, similar to the employees who engage in threatening behavior, will talk about their dissatisfaction and, in some cases, brag about their intended activity. Many ask why someone who has the intent to violate rules and engage in criminal behavior would tell anyone about the intended activity. The answer rests with one of the key motivations to commit such acts: the never-ending need for validation and recognition.

In most cases of suspected espionage, early intervention significantly minimized the damage to U.S. interests. Therefore, personnel and physical security become the primary mechanism for prevention, detection, and early intervention. Given that behavior that violates corporate policy is the end result of a pattern of behavior that moves along a continuum from ideation to action, attention to organizational practices and violations is the best early detection method.

In the past, naïve and uninformed people believed that spies and those who engaged in espionage were "crazy." This was found to be far from the truth. In many cases, individuals who suffer from a significant mental disorder such as schizophrenia or a psychosis are rarely, if ever, granted a clearance. In some instances, they may become mentally ill while cleared for access, but the resulting bizarre and overt nature of their behavior ends their eligibility. More recently, individuals who suffer from bipolar disorder, a psychiatric condition that can have a psychotic component, may retain access in a government setting if they demonstrate compliance with treatment and insight into their illness. Similar to alcoholism and gambling, these conditions, when known, can be managed and monitored. These problems offer a greater challenge in the civilian workplace. They clearly require a team approach that brings a variety of expertise to bear on the problem. Issues related to the Americans with Disabilities Act and other labor practice rights, responsibilities, and risk must be carefully reviewed and considered.

Individuals with certain types of personality disorders are considered to create greater risk of insider threat than anyone identified as mentally ill. A personality disorder is best characterized as a pattern of behavior that is marginally adaptable, leading to conflicts in relationships, difficulties at work, and periodic emotional shifts. As a result of poor judgment, behavior can become self-defeating and sometimes self-destructive. Several personality disorders are considered to be of greatest risk in the insider threat arena.

Psychopathic Personality

The psychopathic personality style is frequently identified as the constellation of character traits of greatest risk for insider threat. This personality disorder is characterized as having an orientation to here and now, with little, if any, interest in the future and no interest in learning from the past. These individuals are very self-serving and seek immediate gratification. They have very little, if any, frustration tolerance or ability to delay gratification. The hallmark indicator of the disorder is the lack of guilt or remorse. They have little, if any, capacity to develop commitments to anyone or anything. Their ability to develop any degree of loyalty is seriously compromised. As a result, they do not view rules as having any relevance for them and the limits set for others do not apply to them. The psychopathic personality type is manipulative and exploitative in relationships and situations to serve his or her own immediate needs. Self-aggrandizement and a strong sense of entitlement are hallmark characteristics. Finally, the individual displaying a psychopathic personality may engage in criminal conduct and, as a result, become involved with the criminal justice system (Kernberg, 1992).

Narcissistic Personality

The second personality disorder that is associated with the trusted insider is the narcissistic personality disorder. The narcissistic personality disorder is akin to the psychopathic personality based on some similar traits; however, this individual in many cases can make a commitment and develop a loyalty. In fact, many very successful overachieving executives are clearly narcissistic personalities. The critical concern with the narcissistic personalities is the degree of entitlement, sensitivity to criticism, and the self-aggrandized view that leaves them vulnerable to acting out against an organization or a person when they feel that they have been unfairly criticized. This often leads to a feeling of being devalued. Some individuals with this type of personality will seek out other sources for validation and affirmation to compensate for the source or organization they feel has undervalued them. As a result of feeling entitled, they may think that rules do not apply to them and they may engage in behavior that violates the organization's rules and policies.

The narcissistic personality is characterized as often grandiose and self-serving. Unlike the psychopathic personality, these individuals are achievement oriented and can develop some limited attachments. They have some impulse control and are able to tolerate a degree of frustration in regard to having their needs immediately met. They can learn from past mistakes. They have motivation to pursue future goals. Often, these individuals are competitive and aggressive in their pursuit of success. Therefore, in cases where they perceive to be valued by the system and can use the system to sustain their self-aggrandized view of themselves, they can become very successful. When these individuals believe that their talents are not appreciated, they can be quick to seek out other sources of validation. Depending on what life crisis they are experiencing, the choice can result in a serious potential for security violations or espionage. As a result of these individuals' positive disposition or success and the frequent absence of any disqualifying adjudicative criteria, they are frequently cleared and given access.

THE DEVELOPMENT OF A PROBLEM

As noted earlier, trusted insiders begin to develop problems with the organization that they perceive as unsolvable. In many instances, individuals may consider committing espionage (e.g., "if I shared this information, I would be viewed as important or gain some financial reward"), but never carry out an action. Others engage in behavior that is more consistent with someone who can move along a continuum of idea to action. They may explore and rehearse how it might be done with a positive outcome, leading on occasion to an act of betrayal or espionage. For individuals to choose to take some action, three critical factors need to align. These factors include an individual predisposition, a life crisis, and access to secrets. A life crisis, such as being devalued or rejected, may act as a triggering condition. Individuals begin to project blame onto a situation for their own shortcomings to avoid feeling inadequate, and begin to make decisions that promote self-protection and self-preservation.

Identifying actions that demonstrate a pattern of behavior where commitment, loyalty, and good judgment are lacking becomes the cornerstone of prevention and the management of potential risk. Therefore, the prevention of a betrayal of trust is best managed by identifying risk patterns of behavior before an individual is given access. The patterns should then be monitored and management should be alerted if changes occur. This allows managers to determine what is going wrong in the relationship before any act of betrayal has occurred. Leadership may, at times, undervalue an employee's ongoing contributions and sacrifices creating the fertile ground for exploitation by others who would like to use the employee to gain access to proprietary information and assets.

Results of espionage studies, although anecdotal, revealed many interesting observations and insight regarding espionage applicable to organizations and the insider threat. These findings were instrumental in contributing to personnel security policy development, altering investigative strategies, and fine tuning adjudicative standards. Although these studies did not generate a profile of a "spy," they identified patterns of behavior that could be recognized early in either a background or security investigation. They generated information that could be helpful to professionals who needed to ask insightful questions about behavior and activity they were either observing or that was being reported. In all cases, the subjects were found to be volunteers. Despite earlier beliefs, no real penetrations or initial recruitment took place. Individuals were recruited after they had volunteered to commit espionage.

The following are some key observations from espionage studies that shed light on insider threats:

- At-risk insiders did not seek employment or a clearance with the intent to commit espionage.
- There was no single motivator.
- The insiders acted to fulfill complex personal needs regardless of whether they were just volunteers or were later recruited.
- The insiders were not mentally ill.
- Personality characteristics were present that could have been identified by further assessment. The personality styles were termed self-indulgent, exploiter, overcompensating, and inadequate.
- Insiders saw themselves as special and deserving, but were dissatisfied and felt unfulfilled.
- Insiders engage in several defensive maneuvers within themselves to sustain their own actions through self-deception. They rationalize and minimize their acts and frequently hold others, including the organizations, partially responsible for their actions.
- Insiders felt no guilt about their betrayal because they did not perceive that they were betraying national security.
- Monies received are spent, not saved; yet unusual affluence is not overly apparent.
- Insiders perceive great ease in conducting their activities.
- Insiders tend to tell trusted others of what they are doing. Sometimes this is for support; often it is to impress and involve.

Just as many negative factors identify with potential information risk and possible espionage, a number of mediating factors balance some risk indicators. The factors that appear to be most useful in balancing identified risk factors include evidence of long-term commitments or relationships and the capacity for loyalty and social consciousness.

This information can be useful for investigators to assess at-risk behavior in interviews. These data highlight the necessity to develop and include relevant data in background reports where these are routinely done. It emphasizes the need to identify potential risk behaviors and assess mediating factors.

Finally, it helps direct security managers to identify appropriate points for consultation with mental health or counterintelligence professionals. In regard to the behavioral indicators noted previously, a number of personality types associated with risk factors can be useful in thinking and formulating an understanding of potential security risks.

BEHAVIORAL INDICATORS OF INSIDER RISK

Personality Styles:

- *Self-centered:* self-important, cocky, belligerent, resentful, pretentious, overly dramatic. Constantly seeks recognition and admiration beyond merit.
- *Arrogant:* views rules as naïve or inapplicable to self. Reveals a careless disregard for personal integrity and an indifference to the rights of others.
- *Adventurous:* seems attracted to risk, danger, and harm. Impulsive, restless, impetuous. Displays intolerance of boredom and inactivity. Considers conventional lifestyles beneath him or her. Seductively flirtatious and exhibitionist. Impersonal, trivial sex life.
- *Manipulative:* shamelessly takes others for granted and uses them to enhance self and indulge desires. Disregard for obligations.
- *Cold:* aware of, but indifferent to, feelings of others. Not empathetic.
- *Grandiose:* has an undisciplined imagination and exhibits a preoccupation with immature fantasies of success, beauty, or love. Is minimally constrained by objective reality. Often lies to redeem self-illusions. Speech characterized by exaggeration and hyperbole. Preoccupied with fantasies of success/power. Overly concerned with physical appearance.
- *Self-deceptive:* facile in devising plausible reasons to justify self-centered and socially inconsiderate behaviors. Offers alibis to place oneself in the best possible light. Failure to believe his or her behavior will be punished.
- *Defensive:* reacts to criticism with anger/rage. Overreacts to constructive criticism.

In these types of situations teams need to expect that the data provided to them will explore a number of areas. Many of these overlap with data one would expect threat assessment experts to address in other threat situations. Four are more specific: reactive to criticism, entitlement, grandiosity, and manipulative. These questions are helpful to consider when the team is reviewing at-risk behavior for insider threats.

Reactive to Criticism

General

- How is the individual as a supervisee?
- How does the individual handle criticism?
- Has the individual ever overreacted to criticism or failure (e.g., petulance, temper tantrums, extensive written appeals)?
- Would you describe the individual as "thin skinned" or "touchy"?

Auxiliary

- Does the individual blame conditions or someone else for a mistake that was truly his or her doing?
- Does the individual seem to feel persecuted by others in the office?
- Has the individual ever been unresponsive to criticism?
- Are his or her reactions what you would expect?
- Is the individual inclined to hold grudges?

Entitlement

General

- Does the individual tend to press the limits of rules, procedures, and regulations?
- Can you think of a time when the individual seemed to ignore or neglect an obligation?
- Have you known the individual to bend or break the rules when it served his or her interests?

Auxiliary

- Does the individual break a number of minor rules?
- Is he or she frequently tardy?
- Does he or she consistently leave earlier than is reasonable?
- Is there any suspicion that the individual abuses leave policies?
- Is the individual inclined to disregard proper security procedures?
- Does the individual openly derogate security procedures?
- Does he or she not seem to appreciate the purpose of rules or procedures?
- Does the individual seem to ask more favors than he or she is willing to return?
- Does the individual seem to be particularly envious of others?

Grandiosity

General

- How does the individual rate his or her own work performance and abilities?
- Are these significantly higher than you (the supervisor) see the performance?
- Does this individual like to be the center of attention?

Auxiliary

- Does he or she think his or her performance is much better than it actually is?
- Does the individual anticipate promotions that are probably not forthcoming?
- Does the individual seem to feel that he or she can achieve great things, such as being very successful, powerful, or brilliant?
- Are his or her expectations unrealistic?
- Does he or she tend to exaggerate accomplishments and talents?
- Does the individual seem to think he or she has special talents or abilities that others have not recognized?

- Does the individual seem to actively seek out recognition or praise more than others under your supervision or in your experience?
- Does the individual seem to be overly concerned with his or her physical appearance?

Unstable Relationships

General

- How does the individual get along with his or her co-workers?
- Does the individual have friends in the workplace?
- Is the individual inclined to be argumentative?

Auxiliary

- Do the individual's feelings toward others seem to change in that sometimes he or she loves or greatly admires someone and then hates or feels terribly disappointed by the same person at another time?
- Would you describe the individual as warm or sympathetic?
- Does the individual tend to be insensitive to others?
- Is the individual inclined to be aloof, withdrawn, or uninvolved?
- Does the individual seem to view the world as made up of friends and enemies?

Manipulative

General

- Does the individual tend to be insensitive to others?
- Would you describe him or her as "driven"?
- Would you describe the individual as a good "team player"?
- Is the individual inclined to take advantage of, or use, others to get what he or she wants?

Auxiliary

- Is he or she inclined to get involved in office conflicts?
- Have you ever known the individual to exploit others?

- Does the individual seem to be overly ambitious in seeking relationships with others who are in positions of power?
- Does he or she seek those relationships to the detriment of work relationships or work performance?
- Is the individual inclined to be deceptive?
- Does he or she seem unconcerned with the rights or welfare of others?

APPLICATIONS

This information can be applied to insider threats in a number of ways. If the assumption is that insiders pose the most significant threat, then adjudication becomes a critical preventive intervention. Granting eligibility can be one of the most significant risk management decisions made in regard to counterintelligence. The following are risk control actions regarding the application of this information in the information security arena.

- Background investigator training/outsourcing with appropriate fair credit reporting procedures
- Team training
- Reporting guidelines and awareness training
- Investigative enhancement/internal or external resources
- Interview training/external resource identified

Several critical factors should be hallmark signs to potential risk. Identification of these risk factors should almost always result in further exploration and review, which may lead to a full-blown threat assessment and risk control actions where the employee has access to sensitive insider information. Not one of these factors alone is sufficient. Teams need to use expert resources in evaluating the need to move forward in a threat assessment mode or whether the issues are best managed in the usual human resources (management) channels.

- Slow to learn consequences
- Low self-esteem
- Immature
- Thrill seeking, impulsive, easily bored
- Vengeful, vindictive
- Behaviors indicative of emotional distress

- Sense of entitlement
- Rules do not apply
- Unrealistic view of self or achievements
- Manipulative
- Resentful of authority—stretches limits
- Self-deceptive

Suitability factors also play a critical role in the decision-making process for eligibility. The following factors should be reviewed in individuals who hold positions with access to information or technologies vital to the organization. Each case needs to be individually reviewed by the team where risk has become an issue. These factors are specifically identified in a preemployment screening process. All preemployment procedures need to be reviewed by labor counsel.

- Alcohol, substance abuse
- Financial issues
- Criminal behavior (including juvenile)
- Workplace performance or behavior (previous)
- Compulsive or excessive gambling
- Repeated policy violations (rules do not apply)

Managers with ongoing insider threat concerns need to consider the following behaviors as signals that a more in-depth review may be necessary:

- Unexplained affluence
- Falsification of destruction certificates
- Undue curiosity
- Extremely knowledgeable/suspicious of others
- Taking home sensitive materials in violation of policy
- Clandestine behavior
- Unexplained absences
- Attempt to gain new access
- Avoidance of security scrutiny
- Foreign bank accounts
- Approaching others for information outside of job responsibility
- Unusual work hours are present in areas with access to sensitive material
- Computer access violations

We are confronted with an ever-changing world. The emerging threat continues to change with the volatility of political and economic situations throughout the world. These changes will continue to challenge us in regard to risk management of sensitive organizational information and technologies. Organizations have become more aggressive in collecting sensitive and proprietary information. Considerable effort is exerted recruiting and co-opting scientists and engineers for technical and proprietary information. The threat through the targeting and use of information systems continues to be a main focus. Although the vehicle may be the computer, the behavior is the product of human beings. The work of people such as Bottom and Gallati (1984, 2001) blends information on risk and people into a manageable focus for organizations. Significant levels of espionage activity continue against intellectual and industrial secrets, the very assets that an organization may need to survive and thrive.

Policy and procedural violations are early warning signs of possible espionage. Individuals with access to secrets commit the majority of egregious espionage cases. Betrayal was not committed as a result of an impulse; rather it progressed from idea to action. Identifying at-risk behavior, assessing the threat, and taking appropriate risk control acts are the most effective defenses against espionage, and they provide an appropriate level of protection for the organization.

This is a new area to many organizations, one that was thought to be a problem only of the government. The challenge of maintaining and protecting intellectual property, competitive information, and trade secrets has become a new battle that many organizations do not address until a threatening breach has occurred. Managers need to develop procedures to address these issues proactively, fairly, and in accordance with federal and state statutes. The threat management team is one forum uniquely positioned with a range of experience and expertise to guide organizations in risk management.

Chapter 13

Executive Vulnerability

In most cases executives have high visibility both inside and outside their respective companies. Although executives may not be CEOs of Fortune 500 companies, they are vulnerable to any disgruntled employee, shareholder, single issue protestor, or customer/client. The executive is often visible in company communications, media, and other forms of public relations. Web pages may display extensive personal information about an executive. The executive becomes an easy target for both known and unknown individuals who are motivated to communicate a threat or pose a threat.

In certain companies with international exposure, the executive is vulnerable to a number of threats not related directly to the workplace, but arising from being an American executive. These may include abduction, assassination, and politically driven agendas and usually occur in countries experiencing economic and political instability as well as terrorism.

Although most executives will not be subject to these more exotic threats, there are risks for executives close to home. Situations often arise as a result of workplace violence or significant organizational upheaval. The executive receives communicated threats via mail, fax, or electronic communications in response to changes in policy, production, and personnel. When a company downsizes, the CEO can readily become the target of angry and disgruntled employees. When a downturn in the economy occurs and the company stock loses value, an executive can become the target of threatening communications or be placed at risk at stockholder meetings. In companies where strikes occur as a result of contract negotiations, angry, hostile, and disgruntled employees who feel entitled and exploited by management can sometimes bypass their legitimate representatives and take action into their own hands.

When executives receive threatening letters and packages, the executive's administrative assistant often has the first contact. In other cases, approach behavior follows a series of communications as the subject moves from thought to action. A CEO of a large company was threatened at a shareholders meeting when an elderly shareholder, holding what appeared to be a hand grenade, approached the CEO. The hand grenade was a fake; however, it caused considerable emotional distress. Upon review, it was learned that the individual had made several calls to the organization's public relations group complaining about stock losses.

Executives, by the nature of who they are and the jobs they hold, are used to being in control. They are also used to having people respond to them with a degree of urgency and completion so they can move on to the next task. One issue facing the threat management team is the role of the executive.

A unique opportunity exists at the time of entry into the organization of a new executive for the threat assessment team. They should hold a very brief orientation, ten minutes or less, on the role of the threat management team including their role in supporting the executive as well as every other employee in the organization.

It is also important to invite executive staff to permit scenario-based training of the threat management team. Although the team may have several days of training over a range of scenarios, forty-five minutes will give executive staff members an understanding of the complexity of the challenges and the skill of the threat management team in developing risk control actions.

The management representative with decision-making authority needs to be someone other than the individual at risk. If a member of the team is the perceived target, use an alternate on the team. Individuals with emotional involvement in the situation are more likely to have difficulty maintaining a clear focus on the total picture. At some later date there may be concerns about whether they can make impartial decisions based on the available data. The level of emotional distress may be significantly increased by being exposed to initial reports that are incorrect or unsubstantiated.

The most critical issue in the team assessment of executive vulnerability is the evaluation of the executive. The degree to which an executive is inexperienced in the management of threat behavior and the need for risk control actions will influence the team's ability to manage and advise senior management. The executive who is able to take advice from security and other professionals on the threat assessment team is more likely to be safe. The executive who refuses first to listen to security recommendations, and second to follow advice in security precautions and procedures is more likely to be vulnerable.

The team needs to be free to explore a variety of potential scenarios including worst case ones so that risk control actions may be mapped and contingencies provided. The ability of the team to do this with the potential target of the threat present is limited. A second major area of concern is the use of outside consultants, both legal and threat assessment. At times, very sensitive issues related to the organization's business may be involved. Questionable business dealing, mergers in which people were not treated appropriately, and romantic liaisons do happen. The use of outside consultants can insulate the organization and the executive from situations that might impair the ability to conduct the organization's activities. In such cases, legal counsel should take a lead role in directing the threat assessment and risk control actions to establish attorney client privilege and attorney work product status for reviews.

Gaining information control is a critical issue. Threat assessment teams need to establish a system in which correspondence that is likely to be related to the ongoing situation is monitored and screened. Letters that may appear to the average person as benign may provide important insights for the threat assessor. Letters that might have been discarded as "nut letters" should be retained and inventoried as they may contain threat-related information that indicates a serious escalation. Although individuals who communicate threats may only be discharging frustration in an expressive hoax, they may be communicating an organized, fixed, and focused communication that contains an action and time imperative.

A series of risk control actions must be developed for the individual who serves as the gatekeeper. This needs to include accessing help, exit strategies, and training in de-escalation skills. For specific individuals who may try to access the executive, stay away letters should be prepared. Appendix B gives an example of a stay away letter and instructions for preparation.

The purpose of such tools is to disrupt the forward motion of the individual. Such individuals have rehearsed the approach incident over and over again. They have decided on a very linear course of action in which they have mentally rehearsed and planned the outcomes. Unexpected disruption techniques serve to derail the individuals from their carefully constructed paths. They are often left at a loss as to what to do and respond to suggestions in the letter that another path is open to them. The presentation of the letter in the prescribed format also slows the approach process, giving the gatekeeper time to access security support.

Also, the executive's assistants should be kept out of the loop in terms of reading such incoming correspondence. Reading them will only increase anxiety, emotional distress, or anger. If necessary, threat assessment professionals can involve them in a limited review of the content when the meaning or reference is unclear. Last, individuals should be given and requested to use self-sealing plastic bags for any suspicious material. This not only preserves evidence, it inexpensively limits any bioterrorist threat.

Web sites related to the organization should be reviewed for personal data, including pictures. If the communication begins to indicate personal knowledge, it is important to know if this data is publicly available or if this represents an increase in the threat variables because the individual has spent time and energy to obtain this protected information. Travel schedules and public appearances may be readily available on such sites. Shareholder meetings represent a special risk area when times are bad and/or specific ongoing threat communication exists. Special security and crisis communication planning needs to occur. The threat team and the event planners need to establish clear channels of communication and contact points. An odd event during the preparation of an event may mean little to the event planners. When that information is married to an ongoing risk assessment, the need for specific risk control actions may become evident.

Threat teams need to be tied into the public relations and communication system. Media coverage and press releases may be the catalyst for additional threatening communications or actions directed at the organization or the executive who is the symbolic representative of the organization. One client had a predictable pattern that whenever a newspaper or a television or radio program covered some positive activity related to the organization, clippings with comments and detailed pages of the wrongdoing of the executives would appear. Eventually, public relations would notify the threat response teams so they could anticipate additional communications and remind individuals of the appropriate procedures. Media attention to threatening communication can result in a copycat phenomenon in which multiple communications from a variety of sources are received.

In Appendix A is a sample card that the threat assessment team needs to provide to potential targets. Where appropriate and with the individual's consent, cards can also be provided to family members including adolescent children. In extreme cases, such information with consent can be provided to schools. An employee who is at risk for threats, potential violence, or harassment may be subject to contact by the individual of concern. This contact may occur outside of organizational property in places such as malls, restaurants, hotel lobbies, or shopping areas and may result in the need to call for a law enforcement or security response. In such situations, the individual should be able to communicate effectively with the responding patrol or security officer(s). Responding officers are likely to be faced with an individual who attempts to orally communicate, but complicates the chronology of past occurrences. These same officers may well get a similar story from the perpetrator with the roles reversed so that he or she is perceived as the victim. The card will assist the responding officer in quickly determining that the employee or family member is the likely victim and the other individual is the likely perpetrator.

As also mentioned in Chapter 7, we recommend the preparation of plastic laminated cards (large wallet size), which provide the responding officers at a glance with all of the relevant information needed for the police/incident report they will need to prepare later. These cards should be kept on the individual's person or in a wallet, purse, locked glove compartment, gym bag, etc. Legal counsel should review and approve the blank template prior to use.

Side one contains the information about the employee and/or family member. A brief physical description helps identify the individual to the responding officer and reinforces that thought has gone into assessing the threat and putting this risk control action in place.

Side two contains information about the alleged perpetrator. This may vary depending upon the status of the individual: employee on administrative leave, terminated employee, nonemployee. Again, information such as previous/ongoing law enforcement interest in the case is provided. This suggests that others in the law enforcement hierarchy may have an interest in the responding officer's handling of the situation.

In summary, a comprehensive threat assessment is the foundation of an effective intervention (Boim and Smith, 1994). The threat management team needs to develop a series of steps to properly identify threat and develop a risk control plan. Status and profile contribute to both unique vulnerabilities and exposure.

Appendix A

Threat Information Card
for Potential Victims

Person Carrying the Card (Side One)

Full Name:
Physical Description:
Title:
Date of Birth:
Driver's License #:
Office Address:
Office Phone #:
Office Fax #:
Employer Point of Contact:
Phone # for Employer Point of Contact:

. . . .

Employee on Administrative Leave (Side Two)

Full Name:
Physical Description:
Address:
Social Security #:
Date of Birth:
Driver's License #:
Law Enforcement Officer(s) Involved:
Law Enforcement Agency Involved:
Law Enforcement Agency Address:
Law Enforcement Agency Phone #:
Police Report (#)s:
Date Noticed with Regard to Trespass:
TRO (Temporary Restraining Order)/RO (Restraining Order) #:

Terminated Employee (Alternate Side Two)

Full Name:
Physical Description:
Address:
Phone #:
Social Security #:
Date of Birth:
Date of Employment Termination:
Driver's License #:
Law Enforcement Officer(s) Involved:
Law Enforcement Agency Involved:
Law Enforcement Agency Address:
Law Enforcement Agency Phone:
Police Report #(s):
Date Noticed with Regard to Trespass:
TRO/RO #:

. . . .

Third Party or Nonemployee (Alternate Side Two)

Full Name:
Physical Description:
Address:
Phone #:
Social Security #:
Date of Birth:
Driver's License #:
Law Enforcement Officer(s) Involved:
Law Enforcement Agency Involved:
Law Enforcement Agency Address:
Law Enforcement Agency Phone:
Police Report #(s):
Date Noticed with Regard to Trespass:
TRO/RO #:

Appendix B
Stay Away Letter

This is a sample of a letter that may be held at a reception or point of contact area if an employee of concern makes an unanticipated appearance at the work site, or it may be sent to an employee who has no reason to be at the work site. Such letters are intended to encourage the employee to leave the work site now or to stay away from the work site, perhaps permitting contact to occur later under more controlled conditions, e.g., telephone contact. This sample letter is for illustration purposes only. It should be edited/tailored to match the organization's own mission and/or values statement and to the circumstances of the individual case. When completed, the letter should be placed in a standard size A4 envelope and double-taped closed. The A4 envelope should then be placed in a large manila envelope and double-taped closed. The large manila envelope (or envelopes if there are multiple points of reception in the work site) can then be left with the receptionist in case the individual shows up. The time involved in opening the multiple envelopes and reading the somewhat lengthy letter serves to interrupt the forward motion and goal-directed behavior of the individual, allowing the receptionist time to summon assistance if the individual will not comply with the "leave now" requests of the letter.

Date

Mr. John Smith
322 UUYHT
JJJJJJJ. AA 11111

Dear Mr. Smith:

As you are aware, The Organization always strives to produce the best possible entertainment products for our customers [and/or provide the best possible service for our customers], and to try to improve our product [and/or service] as we listen to our customers' feedback. To achieve these goals, XCS works to provide an environment for employees that is safe, healthy, and productive. Further, the Organization knows that each employee's contribution is important, and that the team effort of everyone working together will enable us to meet our goals of service to our customers.

*Our Organization's Mission and Values Statement [or equivalent] expresses these core values, which include the following items [the following is sample text; substitute actual text from company values or handbook statement]:

- The Organization has an obligation to provide a safe and healthy work environment in which all employees are able to fulfill their individual contribution to our service goals.
- All employees, in turn, have an obligation to perform their work in a coordinated and cooperative manner.
- Each employee is a team member, and is expected to treat other team member employees with mutual respect and concern.
- All employees are encouraged to increase their skills and knowledge in areas needed by the Organization, and apply for advancement within the Organization based upon skill and past job performance.
- The Organization is a member of our community, and contributes to civic improvement and charity to others in need.

At present, you are on a _____ [insert the proper word for his current status] status. For important legal and personnel reasons, former employees need to follow the conditions of their _____ status. Contact needs to occur through the HR [Personnel] Department representative by phone [insert number] or by mail, and you should not come to work sites.

[This paragraph may be used as appropriate] I am not available to talk with you at this time. Unfortunately, it is not possible to see you today. However, I recognize the importance of hearing your requests or concerns, and I do want to be available to do that. To accomplish this, I will call you at your home telephone number tomorrow at 9:00 a.m. Please be available at your home telephone number at that time for my call.

**It would be best at this point for you to follow the conditions of your leave status and leave the work site now. I will be able to talk with you at our scheduled time tomorrow.

---

*When the employee has been terminated insert a replacement paragraph that states the specifics of the termination here.

**When the employee has been terminated and it is not desirable to have the employee return to the work site, insert the following paragraph instead: "It has been our experience that on occasion some former employees begin to feel that they may be able to effectively appeal their employment status change by appearing at the office even though the terms of the status change clearly state that this is not a useful idea. This letter and your prior telephone conversations with company executives indicate the best businesslike procedures for further contact with the company. Approaching the situation in that manner is the best way for us to continue to maintain a businesslike approach with each other in a situation as admittedly difficult as the present one."

Thank you for your cooperation and assistance in following the conditions of your leave status.

Sincerely,

Fred
HR Representative

Appendix C

Behavior Improvement Plan

The following text is a draft of a behavior improvement plan submitted for review and editing by labor law counsel and human resources.

DRAFT BEHAVIOR IMPROVEMENT PLAN
FOR MR. JOHN SMITH

As we have previously discussed, ABC Corporation is committed to providing all employees with a safe and positive work environment, where each individual is allowed to contribute toward the successful attainment of our business goals. As a result of a recent review of your on-site work behavior, the organization has determined that your behavior was unacceptable and in violation of organization policy.

As a part of your return to work, the following behavior improvement plan between you and the organization needs to be implemented:

1. Mr. John Smith acknowledges that the organization has a reasonable concern in not wanting any threats of violence or violence to take place at the workplace.
2. Mr. John Smith acknowledges having made statements that referenced acts of potential violence in the workplace. Such statements are unacceptable, have no purpose in a business environment, and are in violation of organization policy.
3. Mr. John Smith acknowledges that such statements concerning acts of potential violence may reasonably cause concern, intimidation, and emotional distress in co-workers. Such statements may also reasonably result in employee time and energy being diverted from our business purposes in order to deal with such statements' impact upon others at the workplace.
4. Mr. John Smith agrees he will not make any further statements or references to acts of potential violence in the workplace.

5. Mr. John Smith agrees that he will, at all times, conduct himself in a businesslike and professional manner.
6. Mr. John Smith agrees to maintain open communication between himself, his organization supervisors, and the Human Resources Department representatives. Mr. Smith agrees to openly and directly discuss with his organization supervisors and/or Human Resources Department representatives:
   - Any concern or unresolved issue pertaining to his work
   - This workplace behavior improvement plan
   - The events now or in the future that might prevent his completion of this behavior improvement plan
7. As a part of this workplace behavior improvement plan, Mr. John Smith agrees to participate in a counseling program to address his behavior at work, the impact of his behavior upon co-workers, and his plan for preventing the future reoccurrence of this behavior. The organization will arrange for Mr. Smith to meet with a qualified professional through our Employee Assistance Program (EAP). The Organization will notify Mr. Smith of the designated counselor and the telephone number where the counselor can be contacted. Mr. Smith agrees that it will be his personal responsibility and obligation to make telephone contact with this counselor on the day that the referral to the counselor is made. Mr. Smith agrees to make and keep his first appointment with this counselor within five working days of receiving the referral. It is expected that Mr. Smith will meet with this counselor a total of approximately five times. The organization will be notified by the counselor when this workplace behavior counseling program has been completed.
8. Firearms, guns, and weapons of any kind are not allowed in the workplace. Mr. John Smith agrees that he will observe and abide by all applicable organization, state, and local laws, ordinances, and regulations pertaining to the possession and use of firearms/weapons. Mr. Smith further agrees that he will not have any such weapons while on organization work time, or on any organization facility or property (including organization vehicles). Violation of these conditions may result in immediate termination of employment.
9. While at work, if Mr. John Smith should feel the beginnings of loss of temper control, impulse control, thoughts to injure any person, or thoughts to make statements of past or future violence against others, he shall notify a member of the organization's supervisory and/or management team and shall leave the work site immediately. Mr. Smith further agrees that he will then proceed to contact his treating psychiatrist or psychologist and/or the hospital emergency room des-

ignated by his treating professional and request psychological assistance.

10. While at work, if a member of the organization's supervisory and/or management team shall determine that Mr. John Smith's statements and/or behavior are cause for concern about his loss of impulse control, loss of temper control, and/or potential physical risk to other workers, a member of the supervisory and/or management team may direct Mr. Smith to leave the work site immediately. Mr. Smith agrees to comply immediately with this instruction. Mr. Smith further agrees that he will then contact his treating psychiatrist and/or the hospital emergency room designated by his treating psychiatrist and request psychological assistance.

11. Mr. John Smith acknowledges that failure to comply with this agreement may result in disciplinary action, up to and including termination of employment.

Acknowledged and agreed to by:

_____

Mr. John Smith

_____

Organization Representative

Appendix D

Health Provider Questions

The following questions are useful in determining the level of confidence that a treating professional is willing to assign to an opinion. Qualifying statements, e.g., "I can't answer that type of question," "I don't know how to be that precise," "This is not a situation where you could say that," "I can't answer that, but it is my professional opinion that . . . ," "He has not done anything violent or threatening since I began seeing him," "It is my medical opinion that . . . ," "I can say that to a medical certainty," (which usually means more likely than not or probable, etc.) by the professional indicate caution by the employer in relying upon the opinion.

Please note that some of these questions may not be appropriate in your legal judgment based on the facts of the case and current law or regulations. Please delete or alter any that you feel are not legally appropriate. You may need to address release of information issues based on the rules in your jurisdiction.

Our organization is trying to determine if risk exists in our role as employer. The information you provide will be used by the appropriate persons (specify) in the review process along with other independently gathered information. Please do not provide diagnosis or medical/psychological treatment information. If you feel you need to provide that information, please do so separately and address such information to (specify). In responding to the following questions, please provide direct quotes from the employee as much as possible. If you cannot recall the employee's direct quotes and/or such quotes are not in your clinical notes, please so indicate.

1. Please state what the employee has related to you regarding the chronological sequence of events that led to the administrative leave.
2. Do you have any independent or third party verification of the sequence of events that the employee related to you?
3. Please state what the employee has related to you about his thoughts and judgment process at each point in the chronological sequence of events that led to suspension or employment action.

4. Do you have any independent or third party verification as to whether the employee has related a different account of his thought and judgment process as noted in number 3?

5. Please state what the employee has related to you regarding his understanding of the impact of his behavior upon his co-workers and supervisors.

6. Please state what the employee has related to you regarding his understanding of the need of an employer to maintain a safe workplace and to assess behavior that may present a risk to that workplace.

7. Please state the violence history of this employee, including any contact with law enforcement.

8. Are there any company employees that are at risk? Circumstances?

9. Please state your percent degree of certainty that this employee will not engage in any future violent acts.

10. Please state your percent degree of certainty that this employee will not engage in any threats of violence or acts of intimidation.

11. Please state what the employee has related to you about the circumstances under which he thinks it would be likely that he would engage in a future violent act.

12. Please state what the employee has related to you about the circumstances under which it would be likely that he would engage in a future threat or act of intimidation.

13. Please state what the employee has related to you about specific plan(s) for preventing future violent behavior, threats, and/or acts of intimidation.

14. Please state what the employee has related to you about the circumstances under which it would be likely that preventing future violent behavior, threats, and/or acts of intimidation would not work.

15. If you believe that this employee should return to work, please state if this is an unqualified recommendation, or if you are recommending a work environment accommodation. Please be specific about such a recommendation and why it would qualify as a reasonable accommodation.

16. Please state your understanding of employer obligations under the ADA and the Rehabilitation Act of 1974 as it relates to violent behavior and/or threats.

Appendix E

Announcement to Employees Regarding Security Changes

This is a sample announcement for supervisors to utilize when informing employees of changes in the level of security at the work site. This sample letter is for illustration purposes only. It should be (1) edited/tailored to match the organization's own mission and/or values statement and to the circumstances of the individual case, and (2) reviewed by corporate labor law counsel.

ABC Corporation always strives to produce the best products possible for our customers [and/or provide the best possible service for our customers], and to try to improve our product [and/or service] as we listen to our customers' feedback. To achieve these goals, ABC Corporation works to provide an environment for employees that is safe, healthy, and productive. Accordingly, as you may have noticed, we have made some changes in our security procedures here at the work site. These changes include [insert specific changes, such as reduced entry and exit points, card access, sign in–sign out requirements, parking lot escorts, badging guests, employee badge checking, increased uniformed security presence, increased nonuniformed security presence, etc.]. Changes in security are a part of living in these times, in airports, public buildings, even in our homes. As a company, we need to remember that we are not immune from theft, property loss, or physical risk to our employees. Consequently, the company will, on occasion, use different security procedures in an effort to determine how best to provide a safe work site.*

ABC Corporation fully understands that any new security procedures can change our daily work patterns, consume our time, or just annoy us. The

---

*Alternate if risk or threat is known by employees: "From time to time, the company becomes aware of comments or statements that are not consistent with our harassment free/safe workplace policy. As you are aware, all such comments, whether made intentionally or in jest, are taken seriously and thoroughly investigated. Pending the completion of such an investigation, the company may alter security arrangements."

company also recognizes that any new security procedures can serve as a reminder that the world we live and work in today may not be as safe a place as the world of our parents and grandparents. As these new security procedures are implemented, your patience and your cooperation will be much appreciated.

Bibliography

Barish, R. C. (2001). Legislation and regulations addressing workplace violence in the United States and British Columbia. *American Journal of Preventive Medicine, 20,* 149-154.

Baron, R. A. and Neuman, J. H. (1996). Workplace violence and workplace aggression: Evidence on their relative frequency and potential causes. *Aggressive Behavior, 22,* 161-173.

Barrett, K. E., Riggar, T. F., and Flowers, C. R. (1997). Violence in the workplace: Examining the risk. *Journal of Rehabilitation Administration, 21,* 95-114.

Barrett, K. E., Riggar, T. F., and Flowers, C. R. (1997). Violence in the workplace: Preparing for the age of rage. *Journal of Rehabilitation Administration, 21,* 171-188.

Baxter, V. and Margavio, A. (1996). Assaultive violence in the U.S. post office. *Work and Occupations, 23,* 277-296.

Beck, J. C. and Schouten, R. (2000). Workplace violence and psychiatric practice. *Bulletin of the Menninger Clinic, 64,* 36-48.

Bell, C. C. (2000). *Psychiatric aspects of violence: Issues in prevention and treatment.* San Francisco, CA: Jossey-Bass.

Berg, A. Z., Bell, C. C., and Tupin, J. (2000). Clinician safety: Assessing and managing the violent patient. In Bell, Carl C. (Ed.), *Psychiatric aspects of violence: Issues in prevention and treatment* (pp. 9-29). San Francisco, CA: Jossey-Bass.

Bhui, K., Outhwaite, J., Adzinku, F., Dixon, P., McGabhann, L., Pereira, S., and Strathdee, G. (2001). Implementing clinical practice guidelines on the management of imminent violence on two acute psychiatric in-patient units. *Journal of Mental Health (UK), 10,* 559-569.

BLS (1995). *National census of fatal occupational injuries, 1994.* Washington, DC: U.S. Department of Labor, Bureau of Labor Statistics.

Boim, I. and Smith, K. (1994). Detecting weak links in executive armor. *Security Management, 38*(2), 50.

Boon, J. C. W. and Sheridan, L. (2001). Stalker typologies: A law enforcement perspective. *Journal of Threat Assessment, 1,* 75-79.

Borum, R. (1996). Improving the clinical practice of violence risk assessment: Technology, guidelines and training. *American Psychologist, 51,* 945-948.

Borum, R. (2000). Assessing violence risk among youth. *Journal of Clinical Psychology, 56,* 1263-1288.

Borum, R., Fein, R., Vossekuil, B., and Berglund, J. (1999). Threat assessment: Defining an approach for evaluating risk of target violence. *Behavioral Sciences and the Law, 17,* 323-337.

Bottom, N. R. (2001). Certified counter and information security manager's program (CCISM). Espionage Research Institute (www.espionbusiness.com/faq).

Bottom, N. R. and Gallati, R. (1984). *Industrial espionage*. Stoneham, MA: Butterworth.

Brownell, P. (1996). Domestic violence in the workplace: An emergent issue. *Crisis Intervention and Time-Limited Treatment, 3,* 129-141.

Budd, J. W., Arvey, R. D., and Lawless, P. (1996). Correlates and consequences of workplace violence. *Journal of Occupational Health Psychology, 1,* 197-210.

Buffone, G. W. (2001). Workplace violence: Assessment, prevention, and response. In VandeCreek, L. and Jackson, T. L. (Eds.), *Innovations in clinical practice: A source book* (pp. 371-383). Sarasota, FL: Professional Resource Press.

Bureau of Justice Assistance (1996). *Regional seminar series on developing and implementing antistalking codes (NCJ Publication No. 156836).* Washington, DC: U.S. Department of Justice.

Cannon-Bowers, J. and Salas, E. (1998). *Making decisions under stress.* Washington, DC: American Psychological Association.

Castillo, D. N. (1994). Nonfatal violence in the workplace: Directions for future research. *Questions and answers in lethal and nonlethal violence: Proceedings of the Third Annual Workshop of the Homicide Research Working Group.* Washington, DC: National Institute of Justice.

Citrome, L. and Volavka, J. (2000). Management of violence in schizophrenia. *Psychiatric Annals, 30,* 41-52.

Citrome, L. and Volavka, J. (2001). Aggression and violence in patients with schizophrenia. In Hwang, M. Y. and Bermanzohn, P. C. (Eds.), *Schizophrenia and comorbid conditions: Diagnosis and treatment* (pp. 149-185). Washington, DC: American Psychiatric Press.

Cole, L. L., Grubb, P. L., Sauter, S. L., Swanson, N. G., and Lawless, P. (1997). Psychosocial correlates of harassment, threats and fear of violence in the workplace. *Scandinavian Journal of Work, Environment and Health, 23,* 450-457.

Davis, J. (2001a). The assessment of potential threat: A second look. *Journal of Policy and Criminal Psychology, 16*(1), 1-10.

Davis, J. (2001b). *Stalking crimes and victim protection: Prevention, intervention, threat assessment, and case management.* Boca Raton, FL: CRC Press.

De Clerambault, G. (1942). *Les psychoses passionately.* In *Oeuvre Psychiatrique* (pp. 323-443). Paris: Presses Universitaires de France.

Deirmenjian, J. M. (1999). Stalking in cyberspace. *Journal of the American Academy of Psychiatry and the Law, 27*(3), 407-413.

Doucette-Gates, A., Firestone, R. W., and Firestone, L. A. (1999). Assessing violent thoughts: The relationship between thought processes and violent behavior. *Psychologica Belgica, 39,* 113-134.

Douglas, S. C. and Martinko, M. J. (2001). Exploring the role of individual differences in the prediction of workplace aggression. *Journal of Applied Psychology, 86,* 547-559.

Driscoll, R. J., Worthington, K. A., and Hurrell, J. J. Jr. (1995). Workplace assault: An emerging job stressor. *Consulting Psychology Journal: Practice and Research, 47,* 205-211.

Duhart, D. T. (2001). *Violence in the workplace, 1993-99 (NCJ Publication No. 190076).* Washington, DC: U.S. Department of Justice.

Elliot, R. H. and Jarrett, D. T. (1994). Violence in the workplace: The role of human resource management. *Public Personnel Management, 23,* 287-299.

"Fear and violence in the workplace" (1993). A survey conducted in October 1993 by Northwestern National Life Insurance Company.

Fein, R. and Vossekuil, B. (1998). *Protective intelligence and threat assessment investigations: A guide for state and local law enforcement officials.* Washington, DC: U.S. Department of Justice, Office of Justice Programs, National Institute of Justice.

Fein, R. A., Vossekuil, B., and Holden, G. (1995). *Threat assessment: An approach to prevent targeted violence (NCJ Publication No. 155000).* Washington, DC: U.S. Department of Justice, Office of Justice Programs, National Institute of Justice.

Flannery, R. B. Jr. (1996). Violence in the workplace, 1970-1995: A review of the literature. *Aggression and Violent Behavior, 1,* 57-68.

Flannery, R. B. Jr. (2000). Post-incident crisis intervention: A risk management strategy for preventing workplace violence. *Stress Medicine, 16,* 229-232.

Fletcher, T. A., Brakel, S. J., and Cavanaugh, J. L. (2000). Violence in the workplace: New perspectives in forensic mental health services in the USA. *British Journal of Psychiatry, 176,* 339-344.

Grainger, C. (1996). How controllable is occupational violence? *International Journal of Stress Management, 3,* 17-23.

Heilbrun, K. (1997). Prediction versus management models relevant to risk assessment: The importance of legal decision-making context. *Law and Human Behavior, 21,* 247-359.

Holmes, R. M. (1993). Stalking in America: Types and methods of criminal stalkers. *Journal of Contemporary Criminal Justice, 9,* 317-327.

Jenkins, E. L. (1994). Occupational injury deaths among females: The U.S. experience for the decade 1980 to 1989. *Annals of Epidemiology 4*(2), 146-151.

Jenkins, E. L. (1996). Workplace homicide: Industries and occupations at high risk. *Occupational Medicine State of the Art Reviews 11*(2), 219-225.

Johnson, P. R. and Indvik, J. (1994). Workplace violence: An issue of the nineties. *Public Personnel Management, 23,* 515-523.

Kashani, J. H., Jones, M. R., Bumby, K. M., and Thomas, L. A. (1999). Youth violence: Psychosocial risk factors, treatment, prevention, and recommendations. *Journal of Emotional and Behavioral Disorders, 7,* 200-210.

Kenway, J., Fitzclarence, L., and Hasluck, L. (2000). Toxic shock: Understanding violence against young males in the workplace. *Journal of Men's Studies, 8,* 131-152.

Kernberg, O. (1992). *Aggression in personality disorders and perversions.* New Haven: Yale University Press.

Klein, R. L., Leong, G. B., and Silva, J. A. (1996). Employee sabotage in the workplace: A biopsychosocial model. *Journal of Forensic Sciences, 41,* 52-55.

Kondrasuk, J. N., Moore, H., and Wang, H. (2001). Negligent hiring: The emerging contributor to workplace violence in the public sector. *Public Personnel Management, 30,* 185-195.

Maggio, Mark J. (1996). Keeping the workplace safe: A challenge for managers. *Federal Probation 60*(1), 67-71.

Meloy, J. R. (1988). *The psychopathic mind: Origins, dynamics, and treatment.* Northvale, NJ: Jason Aronson.

Meloy, J. R. (1992). *Violent attachments.* Northvale, NJ: Jason Aronson.

Meloy, J. R. (1998). *The psychology of stalking: Clinical and forensic perspectives.* San Diego, CA: Academic Press.

Meloy, J. R. (2000). *Violence risk and threat assessment: A practical guide for mental health and criminal justice professionals.* San Diego, CA: Specialized Training Services.

Meloy, J. R., Rivers, L., Siegel, L., Gothard, S., Naimark, D., and Nicolini, J. R. (2000). A replication study of obsessional followers and offenders with mental disorders. *Journal of Forensic Sciences, 45,* 147-152.

Miller, L. (1999). Workplace violence: Prevention, response, and recovery. *Psychotherapy: Theory, Research, Practice, Training, 36,* 160-169.

Miller, M. C., Tabakin, R., and Schimmel, J. (2000). Managing risk when risk is greatest. *Harvard Review of Psychiatry, 8,* 154-159.

Mohandie, K. and Hatcher, C. (1999). Suicide and violence risk in law enforcement: Practical guidelines for risk assessment, prevention, and intervention. *Behavioral Sciences and the Law, 17,* 357-376.

Monahan, J. (1985). Evaluating potentially violent persons. In Ewing, C. R. (Ed.), *Psychology, Psychiatry, and the Law* (pp. 9-39). Boca Raton, FL: Professional Resource Exchange.

Monahan, J. and Steadman, H. J. (1994). Toward a rejuvenation of risk assessment research. In Monahan, J. and Steadman, H. J. (Eds.), *Violence and mental disorder: Developments in risk assessment* (pp. 1-17). Chicago, IL: The University of Chicago Press.

Monahan, J. and Steadman, H. J. (1996). Violent storms and violent people: How meteorology can inform risk communication in mental health law. *American Psychologist, 51,* 931-938.

Mossman, D. (1995). Violence prediction, workplace violence, and the mental health expert. *Consulting Psychology Journal: Practice and Research, 47,* 223-233.

Mulvey, E. P. and Cauffman, E. (2001). The inherent limits of predicting school violence. *American Psychologist, 56,* 797-802.

Mulvey, E. P. and Lidz, C. W. (1998). The clinical prediction of violence as a conditional judgment. *Social Psychiatry and Psychiatric Epidemiology, 33,* 107-113.

NIOSH (1992). *Homicide in U.S. workplaces: A strategy for prevention and research.* Morgantown, WV: U.S. Department of Health and Human Services, Public Health Service, Centers for Disease Control, National Institute for Occupational Safety and Health, DHHS (NIOSH) Publication No. 92-103.

NIOSH (1993). *NIOSH Alert: Request for assistance in preventing homicide in the workplace.* Cincinnati, OH: U.S. Department of Health and Human Services, Public Health Service, Centers for Disease Control, National Institute for Occupational Safety and Health, DDHS (NIOSH) Publication No. 93-109.

NIOSH (1995). *National traumatic occupational fatalities (NTOF) surveillance system.* Morgantown, WV: U.S. Department of Health and Human Services, Public Health Service, Centers for Disease Control, National Institute for Occupational Safety and Health, unpublished database.

Office of Justice Program (2001). *Stalking and domestic violence: Report to Congress (NCJ Publication No. 186157).* Washington, DC: US Department of Justice.

Office of Victims of Crime (2002). Help Series Brochures. *Institute of Justice,* www.ojp.usdoj.gov/ovc/publications/infores/help_series/welcome.html.

Otto, R. K. (2000). Assessing and managing violence risk in outpatient settings. *Journal of Clinical Psychology, 56,* 1239-1262.

Palarea, R. E., Zona, M. A., Lane, J. C., and Langhinrichsen-Rohling, J. (1999). The dangerous nature of intimate relationship stalking: Threats, violence, and associated risk factors. *Behavioral Sciences and the Law, 17,* 269-283.

Paul, R. J. and Townsend, J. B. (1998). Violence in the workplace: A review with recommendations. *Employee Responsibilities and Rights Journal, 11,* 1-14.

Prentky, R. A. and Burgess, A. W. (2000). *Forensic management of sexual offenders.* New York: Kluwer Academic.

Purcell, R., Pathe, M., and Mullen, P. E. (2001). A study of women who stalk. *American Journal of Psychiatry, 158,* 2056-2060.

Quinsey, V., Horns, G. T., Rice, M. E., and Cormier, C. A. (1998). *Violent offenders: Appraising and managing risk.* Washington, DC: American Psychological Association.

Resnick, P. J. and Kausch, O. (1995). Violence in the workplace: Role of the consultant. *Consulting Psychology Journal: Practice and Research, 47,* 213-222.

Robinson, L., Littrell, S. H., and Littrell, K. (1999). Managing aggression in schizophrenia. *Journal of the American Psychiatric Nurses Association, 5,* S9-S16.

Rogers, K. and Kelloway, E. K. (1997). Violence at work: Personal and organizational outcomes. *Journal of Occupational Health Psychology, 2,* 63-71.

Rosen, J. (2001). A labor perspective of workplace violence prevention: Identifying research needs. *American Journal of Preventive Medicine, 20,* 161-168.

Ross, L. and Nisbett, R. E. (1991). *The person and the situation: Perspectives of social psychology.* New York: McGraw-Hill.

Runyan, C. W., Zakocs, R. C., and Zwerling, C. (2000). Administrative and behavioral interventions for workplace violence prevention. *American Journal of Preventive Medicine, 18,* 116-127.

Saunders, R. (1998). The legal perspective in stalking. In Meloy, J. R. (Ed.), *The psychology of stalking* (pp. 28-49). San Diego, CA: Academic Press.

Schat, A. C. and Kelloway, E. K. (2000). Effects of perceived control on the outcomes of workplace aggression and violence. *Journal of Occupational Health Psychology, 5,* 386-402.

Scott, C. L. and Resnick, P. (2000). The prediction of violence. In Van Hasselt, V. B. and Hersen, M. (Eds.), *Aggression and violence: An introductory text* (pp. 284-302). Boston: Allyn and Bacon.

Serin, R. C. and Preston, D. L. (2001). Managing and treating violent offenders. In Ashford, J. B., Sales, B. D., and Reid, W. H. (Eds.), *Treating adult and juvenile offenders with special needs* (pp. 249-271). Washington, DC: American Psychiatric Press.

Shafii, M. and Shafii, S. L. (2001). Diagnostic assessment, management, and treatment of children and adolescents with potential for school violence. In Shafii, M. and Shafii, S. L. (Eds.), *School violence: Assessment, management, prevention* (pp. 87-116). Washington, DC: American Psychiatric Press.

Sheldrick, C. (1999). Practitioner review: The assessment and management of risk in adolescents. *Journal of Child Psychology and Psychiatry and Allied Disciplines, 40,* 507-518.

Stone, A. (2000). *Fitness for duty: Principles, methods, and legal issues.* CRC Press: New York.

Swanson, J. W., Swartz, M. S., Borum, R., Hiday, V. A., Wagner, H. R., and Burns, B. J. (2000). Involuntary out-patient commitment and reduction of violent behavior in persons with severe mental illness. *British Journal of Psychiatry, 176,* 324-331.

Tengstroem, A., Hodgins, S., and Kullgren, G. (2001). Men with schizophrenia who behave violently: The usefulness of an early- versus late-start offender typology. *Schizophrenia Bulletin, 27,* 205-218.

Thomson, L. (1999). Clinical management in forensic psychiatry. *Journal of Forensic Psychiatry, 10,* 367-390.

Tiangco, E. O. and Kleiner, B. H. (1999). New developments concerning negligent hiring. *Journal of Workplace Learning, 11,* 16-21.

Tjaden, P. (1997). The crime of stalking: How big is the problem? *Research Preview.* Washington, DC: Institute of Justice.

Tjaden, P. and Thoennes, N. (1998). *Stalking in America: Findings from the National Violence Against Women Survey (NCJ Publication No. 169592).* Washington, DC: U.S. Department of Justice.

Tobin, T. J. (2001). Organizational determinants of violence in the workplace. *Aggression and Violent Behavior, 6,* 91-102.

VandeCreek, L. and Knapp, S. (2000). Risk management and life-threatening patient behaviors. *Journal of Clinical Psychology, 56,* 1335-1351.

VandenBos, G. R. and Bulatao, E. Q. (Eds.) (1996). *Violence on the job: Identifying risks and developing solutions.*

Whitelaw, K. (1996). Fear and dread in cyberspace. *US News and World Report, 121*(18), November 11, p. 50.

Wilkinson, C. W. (2001). Violence prevention at work: A business perspective. *American Journal of Preventive Medicine, 20,* 155-160.

Windau, J. and Toscano, G. (1994). *Workplace homicides in 1992. Compensation and working conditions February 1994.* Washington, DC: U.S. Department of Labor, Bureau of Labor Statistics.

Younger, B. (1994). Violence against women in the workplace. *Employee Assistance Quarterly, 9,* 113-133.

Zona, M. A., Palarea, R. E., and Lane, J. C. (1998). Psychiatric diagnosis and the offender-victim typology of stalking. In Meloy, J. R. (Ed.), *The psychology of stalking: Clinical and forensic perspectives* (pp. 69-84). San Diego, CA: Academic Press.

Zona, M. A., Sharma, K. K., and Lane, J. (1993). A comparative study of erotomanic and obsessional subjects in a forensic sample. *Journal of Forensic Sciences, 38,* 894-903.

Zugelder, M. T., Champagne, P., and Maurer, S. D. (2000). Balancing civil rights with safety at work: Workplace violence and the ADA. *Employee Responsibilities and Rights Journal, 12,* 93-104.

Index